Applications of Graph Theory and Topology in Inorganic Cluster and Coordination Chemistry

R. Bruce King
Regents' Professor
Department of Chemistry
University of Georgia
Athens, Georgia

CRC Press
Boca Raton Ann Arbor London Tokyo

Library of Congress Cataloging-in-Publication Data

King, R. Bruce.
Applications of graph theory and topology in inorganic cluster and coordination chemistry / R. Bruce King.
p. cm.
Includes bibliographical references and index.
ISBN 0-8493-4298-8
1. Chemistry, Inorganic — Mathematics. 2. Coordination compounds. 3. Graph theory. 4. Topology. I. Title.
QD152.5 M38K56 1992
540.2′242′015115—dc20 92-20171
CIP

This book represents information obtained from authentic and highly regarded sources. Reprinted material is quoted with permission, and sources are indicated. A wide variety of references are listed. Every reasonable effort has been made to give reliable data and information, but the author and the publisher cannot assume responsibility for the validity of all materials or for the consequences of their use.

Direct all inquiries to CRC Press, Inc., 2000 Corporate Blvd., N. W., Boca Raton, Florida, 33431.

International Standard Book Number 0-8493-4298-8

Library of Congress Card Number 92-20171
Printed in the United States 1 2 3 4 5 6 7 8 9 0
Printed on acid-free paper

PREFACE

The diverse chemical applications of topology[1] and graph theory[2,3,4,5] have had a significant impact in a number of areas of inorganic chemistry, including structure and bonding in boron cage compounds, metal clusters, coordination compounds, and solid state materials, as well as polyhedral rearrangements. The general purpose of this book is to provide inorganic chemists with a rudimentary knowledge of the areas of topology, graph theory, and related mathematical disciplines necessary to understand their application in inorganic chemistry, with particular emphasis on metal clusters and coordination compounds. The material is presented in a form suitable not only for inorganic research workers but also for use in a special topics course for inorganic graduate students.

Chapters 1 and 2 present the necessary mathematical background in topology and graph theory, as well as relevant areas of group theory. Chapter 3 relates these ideas from topology, graph theory, and group theory to the properties of atomic orbitals so that they can be applied to the understanding of coordination polyhedra. Chapter 4 presents applications of topology and graph theory to boron cage chemistry, showing the relationship of the bonding in three-dimensional boron deltahedra to that in planar polygonal hydrocarbons; Chapter 5 then relates the graph-theory derived treatment of the structure and bonding in cage boranes to numerical computations, thereby providing a foundation for wider application of these methods to inorganic compounds. Chapters 6 through 11 present applications of topology and graph theory to the structure and bonding in other types of inorganic compounds, including metal clusters, solid state materials, metal oxide derivatives, superconductors, icosahedral phases, and carbon cages (fullerenes). Finally Chapter 12 presents an introduction to the applications of topology and graph theory to the dynamics of rearrangements in coordination and cluster polyhedra.

Many of the ideas outlined in this book were developed during the period 1984–1988 as part of a special interdisciplinary research project on chemical applications of topology and graph theory funded by the Office of Naval Research. In this connection, I would like to acknowledge the interest of Dr. David Nelson, then at the Office of Naval Research, in support of this area of

---

[1] R. E. Merrifield and H. E. Simmons, *Topological Methods in Chemistry*, Wiley-Interscience, New York, 1989.

[2] A. T. Balaban, ed., *Chemical Applications of Graph Theory*, Academic Press, London, 1976.

[3] R. B. King, ed., *Chemical Applications of Topology and Graph Theory*, Elsevier, Amsterdam, 1983.

[4] N. Trinajstić, *Chemical Graph Theory*, CRC Press, Boca Raton, FL, 1983.

[5] R. B. King and D. Rouvray, eds., *Graph Theory and Topology in Chemistry*, Elsevier, Amsterdam, 1987.

research. More recently, I would like to acknowledge the help of Prof. Douglas Klein of the Department of Marine Sciences of the Texas A & M University of Galveston, who reviewed the initial draft of the entire manuscript and made a number of useful suggestions. Last, but certainly not least, I would like to acknowledge the patience and cooperation of my wife, Jane, during the various phases of preparation of this book. Without her support of this project, it would never have been completed.

R. Bruce King
Athens, Georgia
April 1992

THE AUTHOR

R. Bruce King was born in Rochester, New Hampshire, graduated from Oberlin College (B. A. 1957), and was an NSF Predoctoral Fellow at Harvard University (Ph.D. 1961). After a year at du Pont and $4^1/_2$ years at the Mellon Institute he joined the faculty of the University of Georgia where he is now Regents' Professor of Chemistry. His research interests range from synthetic organometallic and organophosphorus chemistry to applications of topology and graph theory in inorganic chemistry. He has published approximately 480 research and review articles and is an author or editor of 14 books on organometallic chemistry, inorganic chemistry, chemical applications of topology and graph theory, and chemical conversion and storage of solar energy. He has been the American Regional Editor of the *Journal of Organometallic Chemistry* since 1981. He is the recipient of American Chemical Society Awards in Pure Chemistry (1971) and Inorganic Chemistry (1991).

TABLE OF CONTENTS

Chapter 1
Topology, Graph Theory, and Polyhedra 1
1.1 Topology 1
1.2 Graph Theory 3
1.3 Graph Spectra 6
1.4 Polyhedral Topology 10

Chapter 2
Symmetry and Group Theory 17
2.1 Symmetry Operations 17
2.2 Symmetry Point Groups 18
2.3 Group Representations 23
2.4 Symmetry Point Groups as Direct Product Groups 25
2.5 Permutation Groups 27
2.6 Framework Groups 32

Chapter 3
Atomic Orbitals and Coordination Polyhedra 35
3.1 Atomic Orbitals 35
3.2 Hybridization of Atomic Orbitals to Form Coordination Polyhedra 38

Chapter 4
Delocalization in Hydrocarbons and Boranes 47
4.1 Graph Theory and Hückel Theory 47
4.2 Vertex Atoms 50
4.3 Applications to Globally Delocalized Hydrocarbons, Boranes, and Carboranes ... 53
4.4 Electron Rich Polyhedral Systems 61
4.5 Tensor Surface Harmonic Theory 67

Chapter 5
Relationship of Topological to Computational Methods for the Study of Delocalized Boranes ... 71
5.1 Topological Analysis of Computational Results 71
5.2 Applications to Octahedral and Icosahedral Boranes 73

Chapter 6
Molecular and Ionic Metal Carbonyl Clusters 81
6.1 Metal Carbonyl Vertices in Metal Clusters 81
6.2 Electron Counting in Alternative Theories for Metal Carbonyl Cluster Skeletal Bonding ... 84

6.3 Electron-poor Metal Carbonyl Clusters 85
6.4 Electron Counting in Metal Carbonyl Clusters 87
6.5 Metal Carbonyl Clusters Having Interstitial Atoms 98
6.6 Metal Carbonyl Clusters Having Fused Octahedra 105

Chapter 7
Some Early Transition Metal and Coinage Metal Clusters with Special Features 113
7.1 Early Transition Metal Halide Clusters—a Comparison of Edge-Localized, Face-Localized, and Globally Delocalized Octahedra..113
7.2 Centered Gold Clusters—Non-Spherical Valence Orbital Manifolds..118

Chapter 8
Post-Transition Metal Clusters 125
8.1 Molecular and Ionic Bare Post-Transition Metal Clusters...125
8.2 Polyhedral Gallium and Indium Clusters........................130
8.3 Mercury Vertices in Metal Clusters: Alkali Metal Amalgams..136

Chapter 9
Infinite Solid State Structures with Metal-Metal Interactions 139
9.1 Fusion of Metal Cluster Octahedra into Bulk Metal Structures..139
9.2 Superconductors Having Direct Metal-Metal Bonding........146

Chapter 10
Metal Oxides with Metal-Metal Interactions 161
10.1 Metal-Metal Interactions through Oxygen Atoms.............161
10.2 Delocalization in Early Transition Metal Polyoxometalates: Binodal Orbital Aromaticity 163
10.3 Superconducting Copper Oxides..................................170

Chapter 11
The Icosahedron in Inorganic Chemistry: Boron Allotropes, Icosahedral Quasicrystals, and Carbon Cages 173
11.1 Symmetry of the Icosahedron 173
11.2 Boron Icosahedra in Elemental Boron and Metal Borides 175
11.3 Icosahedral Quasicrystals 182
11.4 Elemental Carbon Cages: "Fullerenes" 188

Chapter 12
Polyhedral Dynamics..195
12.1 Polyhedral Isomerizations..195
12.2 Microscopic Models: Diamond-Square-Diamond Processes in Deltahedra..197
12.3 Macroscopic Models: Topological Representations of Polyhedral Isomerizations..200
12.4 Gale Transformations and Gale Diagrams........................206

Chapter 1

TOPOLOGY, GRAPH THEORY, AND POLYHEDRA

1.1 TOPOLOGY

Topology is the mathematics of neighborhood relationships in space independent of metric distance. In the context of chemical structures, such neighborhood relationships often correspond to the *connectivity* of atoms, i.e., how atoms are joined when all concepts of angles and distances are removed. As we will see in this book, this apparently simple notion of connectivity still includes a substantial amount of chemically significant structural information. This book will focus on specific topological ideas that are useful for understanding the structure, bonding, and properties of inorganic substances. For a more general discussion of topological methods in chemistry, a recent book by Merrifield and Simmons[1] is recommended.

Most of the applications of topology to problems in inorganic chemistry discussed in this book make use of a special type of topological space called a *graph*. The branch of mathematics known as *graph theory* studies the properties of such graphs. In this context a *graph G* is defined as a non-empty set *V* (the "vertices") together with a (possibly empty) set *E* (the "edges"—disjoint from *V*) of two-element subsets of (distinct) elements of *V*.[2] The vertices of a graph thus represent members of the set, and the edges of a graph represent the neighborhood relationship between these members of the set. Graphs may be regarded as a particular type of topological space as defined above.

These ideas can be exemplified by the following graph, which is of chemical interest since it is the 1-skeleton of the tetrahedron. In this context a *1-skeleton* of a polyhedron is defined as the graph consisting of the sets of all of the vertices and edges of the polyhedron in question:

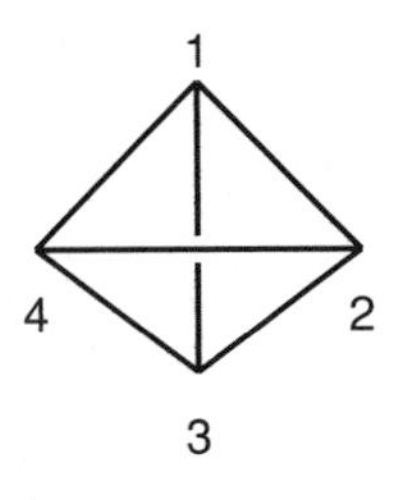

[1]R. E. Merrifield and H. E. Simmons, *Topological Methods in Chemistry*, Wiley-Interscience, New York, 1989.

[2]M. Behzad and G. Chartrand, *Introduction to the Theory of Graphs*, Allyn and Bacon, Boston, 1971, sect.1.1.

This graph has four vertices, which are designated as 1, 2, 3, and 4 and six edges, which may be designated as 12, 13, 14, 23, 24, and 34.

A variety of diverse chemical phenomena can be modelled by topological spaces, as summarized in Table 1–1.

TABLE 1–1
Relationship between Different Topological Models of Chemical Systems

Topological Spaces	Sets	Neighborhood Relationships
Underlying Graphs	Vertices	Edges
Chemical Structure and Bonding	Atoms	Bonding or Antibonding Relationships
Chemical Dynamic Systems	Internal Species (Dynamic Variables)	Dynamic Relationships (Activation or Inhibition)
Topological Representation of Permutational Isomerizations	Isomers	Isomerization Processes

The most obvious possibility is to take the atoms in a molecule as members of the set and bonding or antibonding relationships between the atoms as the neighborhood relationships. Such models are used extensively in this book for the treatment of chemical structure and bonding in inorganic substances. A more general discussion of chemical graph theory is presented in a book by Trinajstić.[3]

Topological methods are also useful for chemical dynamics, since the members of the set can be dynamically active chemical species ("reference reactants" or "internal species") and the neighborhood relationships can be dynamical relationships derivable directly from the underlying chemical reactions.[4,5,6] In addition, *topological representations* have been used to stand for relationships between isomers which can be interconverted by isomerization processes of specific types.[7,8] Some applications of topological methods to polyhedral dynamics are summarized in Chapter 12 of this book.

In general, the methods derived from topology and graph theory for the study of structure and bonding in inorganic species discussed in detail in this book make relatively little use of computation. Nevertheless, they provide information of the following types:

1. Electron counts and shapes of diverse inorganic molecules;

---

[3]N. Trinajstić, *Chemical Graph Theory*, CRC Press, Boca Raton, FL 1983.

[4]R. B. King, *Theor. Chem. Acta,* **56,** 269, 1980.

[5]R. B. King, *J. Theor. Biol.*, **98**, 347, 1982.

[6]R. B. King, *Theor. Chim. Acta*, **63**, 323, 1983.

[7]E. L. Muetterties, *J. Am. Chem. Soc.*, **91**, 1636, 1969.

[8]W. G. Klemperer, *J. Am. Chem. Soc.*, **94**, 6940, 1972.

2. Electron-precise bonding models for inorganic compounds that appear intractable by other methods not requiring heavy computation;
3. Distribution of electrons between internal skeletal bonding and bonding to groups or ligands external to the skeleton;
4. Localized versus delocalized bonding in inorganic structures.

Even though modern computer technology has made major strides in a variety of numerical computations pertinent to the structure and energetics of inorganic substances, information of the above types is often obscure from the results of such computations. Furthermore, the ideas discussed in this book provide information important to the understanding of the structure, bonding, properties, and chemical reactivity of inorganic compounds without depending on a computer.

1.2 GRAPH THEORY

The previous section of this chapter defined a graph as a non-empty set V (the "vertices") together with a (possibly empty) set E (the "edges"—disjoint from V) of two-element subsets of (distinct) elements of V. A *subgraph* of a graph G is a graph H, such that the vertices and edges of H all belong to G. A *complement* of G (denoted by $\overline{G}$) is the graph with the same set of vertices as G, but two vertices in $\overline{G}$ are connected by an edge if and only if they are not connected by an edge in G. There are a number of special types of graphs of interest such as the following:[9]

A. POLYGONAL OR *CYCLIC* OR *CIRCUIT* GRAPHS, C_n

These graphs resemble the regular polygons, i.e.,

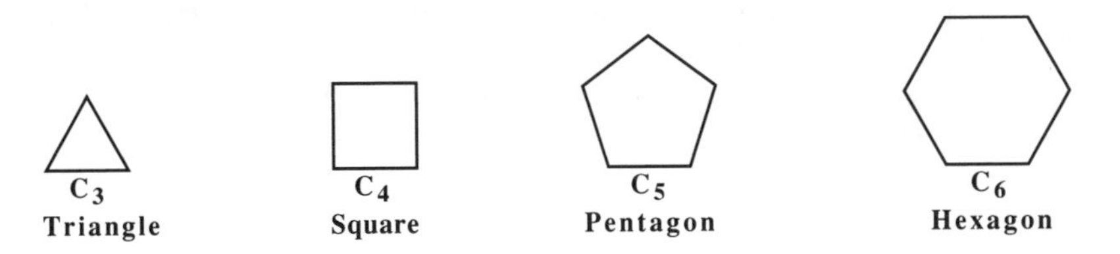

B. COMPLETE GRAPHS, K_n

A complete graph K_n has n vertices and an edge connecting every possible pair of vertices leading to $n(n-1)/2$ edges. A triangle C_3 is also a complete graph K_3 with 3 vertices; note that $3(3–1)/2 = 3$, corresponding to the 3 edges of the triangle. A tetrahedron is the complete graph K_4 on 4 vertices. The K_5 and K_6 graphs are depicted below.

---

[9] L. W. Beinecke and R. J. Wilson, *Selected Topics in Graph Theory*, Academic Press, New York, 1978, Chap. 1.

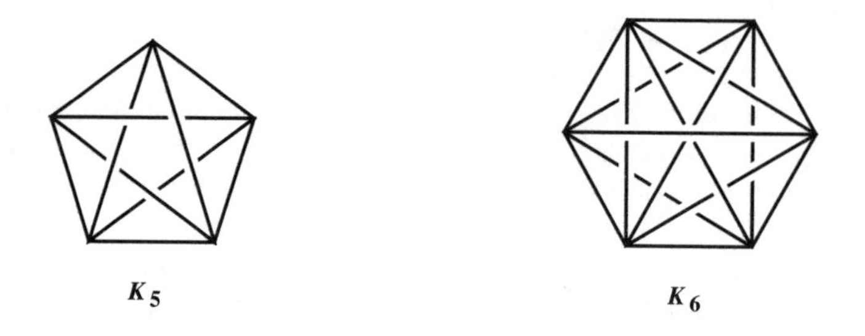

K_5 K_6

C. COMPLETE BIPARTITE GRAPHS, $K_{m,n}$

A complete bipartite graph $K_{m,n}$ is a graph whose vertex-set can be partitioned into two sets having m and n vertices in such a way that every vertex in the first set is adjacent to every vertex in the second set. The square C_4 is also the bipartite graph $K_{2,2}$. Other bipartite graphs of interest are depicted below.

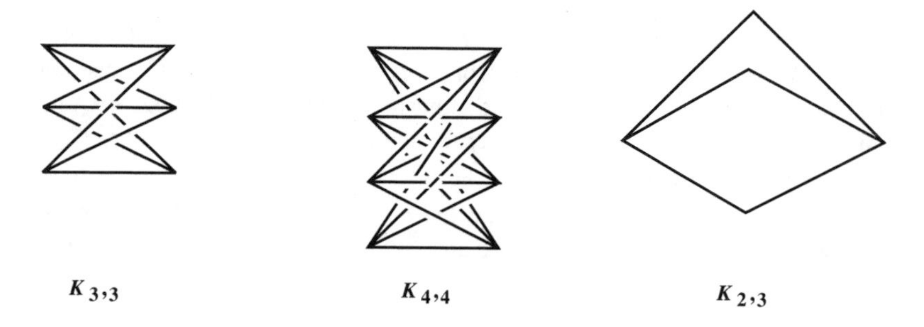

$K_{3,3}$ $K_{4,4}$ $K_{2,3}$

D. LINE GRAPHS

The *line graph L(G)* of G is the graph whose vertices correspond to the edges of G, and where two vertices are joined if and only if the corresponding edges of G are adjacent. For the regular polygons C_n, the line graph is equal to the original graph, i.e., $L(G) = G$. The line graph of the tetrahedron is the octahedron, i.e.,

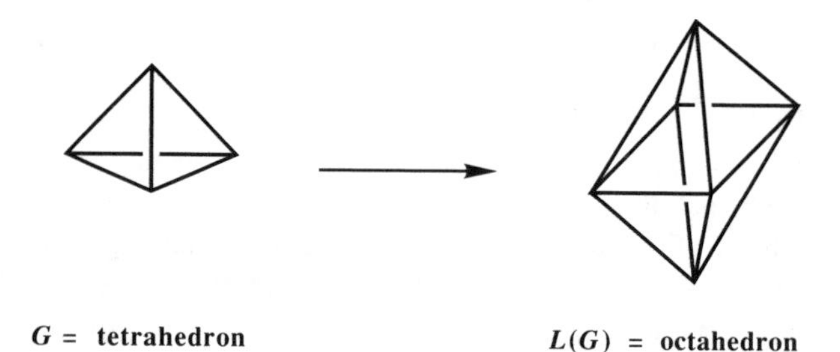

G = tetrahedron $L(G)$ = octahedron

E. PLANAR GRAPHS

A *planar graph* is a graph which can be imbedded in the plane in such a way that no two edges intersect geometrically except at a vertex to which they both are joined. The complete graphs K_n ($n \geq 5$) are *non-planar* graphs, i.e., they cannot be drawn in a plane without edges crossing.

If an edge e of graph G connects vertices v and w, then we can obtain a new graph G' by replacing e by two new edges, namely e' connecting vertex v to vertex z and e'' connecting vertex z to vertex w, by a process illustrated below.

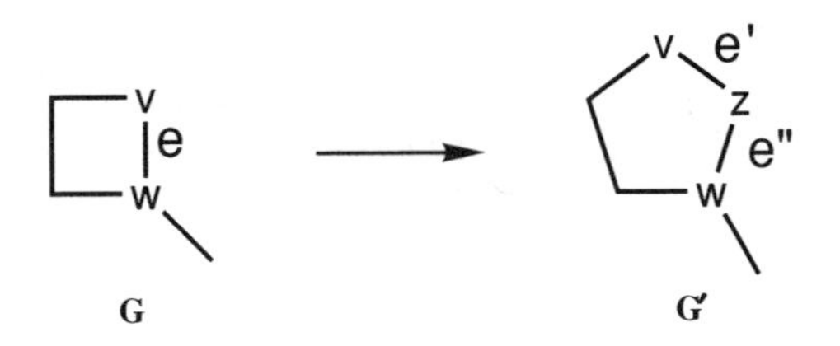

If two graphs can be obtained from the same graph by inserting vertices into its edges in this manner, then these two graphs (e.g., G and G') are called *homeomorphic*. We can also obtain a new graph G'' from G by removing the edge $e = vw$ and superimposing vertex v on vertex w in such a way that the resulting vertex vw is connected to all of those edges (other than e) which were originally connected to either v or w; this is called *contracting the edge e* and can be depicted as follows:

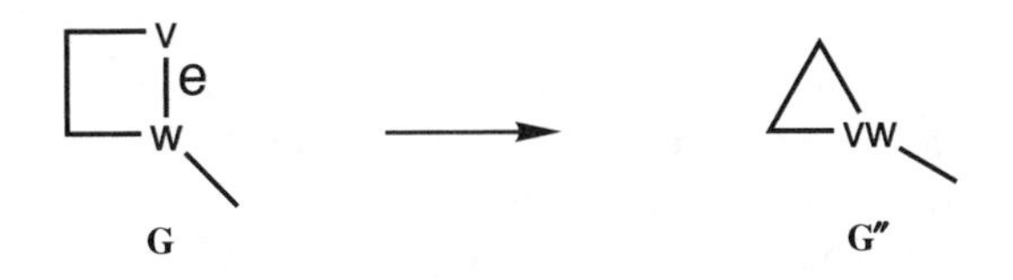

Using these definitions of homeomorphic graphs and contracting a graph, the necessary and sufficient conditions for a graph to be planar can be expressed by *Kuratowski's theorem*,[10] i.e.,

> *A graph G is planar if and only if G has no subgraph homeomorphic to the complete graph K_5 or the complete bipartite graph $K_{3,3}$.*

F. EULERIAN AND HAMILTONIAN GRAPHS

A sequence of edges of the form v_0v_1, $v_1v_2,\ldots,v_{r-1}v_r$ is called a *walk of length r* from v_0 to v_r; if these edges are distinct, the walk is called a *trail* and if the vertices v_0, $v_1,\ldots,v_r$ are also distinct, the walk is called a *path*. A walk or trail is said to be *closed* if $v_0 = v_r$, and a path in which the vertices v_0, $v_1,\ldots,v_r$ are all distinct except for v_0 and v_r (which coincide) is called a *circuit*. A graph G is *connected* if there is a path joining each pair of vertices of G; a graph which is not connected is called *disconnected*. Every

---

[10] K. Kuratowski, *Fund. Math.*, **15**, 271, 1930.

disconnected graph can be split up into a number of maximal connected subgraphs, and these subgraphs are called *components*. A connected graph G is *Eulerian* if it has a trail which includes every edge of G; such a trail is called a *Eulerian trail*. A graph G is *Hamiltonian* if it has a circuit which includes every vertex of G; such a circuit is called a *Hamiltonian circuit*.

1.3 GRAPH SPECTRA

One of the most important properties of a graph from a chemical point of view is its spectrum. Consider first the *adjacency matrix* **A** of a graph G, with v vertices which is a $v \times v$ matrix defined as follows:

$$A_{ij} = \begin{cases} 0 \text{ if i = j} \\ 1 \text{ if i and j are connected by an edge} \\ 0 \text{ if i and j are not connected by an edge} \end{cases} \tag{1–1}$$

The following determinant derived from the adjacency matrix **A** defines the *characteristic polynomial* of G:

$$P_G(x) = |x\mathbf{I} - \mathbf{A}| = \sum_{k=0}^{v} a_k x^{v-k} \tag{1–2}$$

in which **I** is the unit matrix ($I_{ii} = 1$ and $I_{ij} = 0$ for $i \neq j$). The set of v eigenvalues of the adjacency matrix **A** is called the *spectrum* of the corresponding graph G and is obtained by setting the characteristic polynomial to zero and solving the resulting equation, i.e.:

$$P_G(x) = 0 \tag{1–3}$$

Using Equation 1–3 to determine the spectrum of a graph with v vertices thus, in general, requires expanding a $v \times v$ matrix into the corresponding determinant followed by solving an algebraic equation of degree v. In addition, the expansion of a $v \times v$ matrix into the corresponding determinant can be difficult. For these reasons, special procedures have been developed to determine the characteristic polynomial or even the spectrum from a graph G or its adjacency matrix **A**.

By definition the adjacency matrix (Equation 1–1) is not a unique function of the graph because it depends upon the numbering of the vertices. However, different graph numberings merely permute the rows and columns of the adjacency matrix in such a way that the characteristic polynomial $P_G(x)$ (Equation 1–2) remains unchanged. Therefore, the characteristic polynomial and the spectrum are unique properties of the graph, which are independent of the numbering of the vertices.

Despite this uniqueness of the characteristic polynomial and the spectrum of a specific graph, there is *not* a one-to-one correspondence between a graph G and its characteristic polynomial $P_G(x)$. Thus, non-identical graphs can have the same spectrum.[11,12,13,14,15,16,17,18] Such graphs are called *isospectral* graphs. The simplest pair of chemically significant isospectral graphs correspond to *o*-divinylbenzene and 2-phenylbutadiene, i.e.,

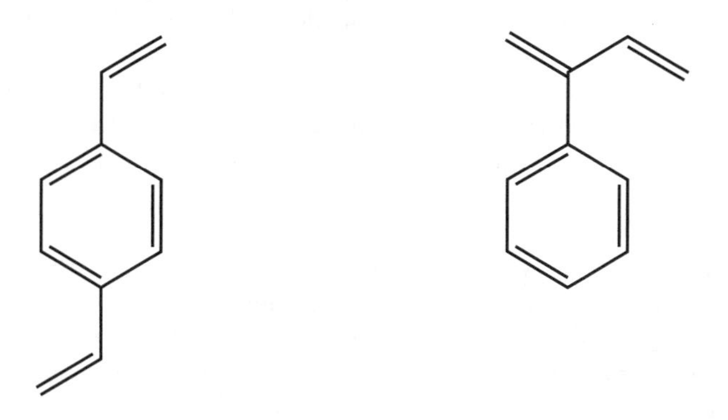

***o*-Divinylbenzene** **2-Phenylbutadiene**

Determination of the spectrum of graph G from its adjacency matrix **A** (Equation 1–1) first requires expansion of the determinant in Equation 1–2 to give the characteristic polynomial. In order to relate the coefficients a_n of the characteristic polynomial to the topology of the graph G, the algorithm of Sachs[19,20] is used. In this connection it is necessary to define a *sesquivalent subgraph* (also called a *Sachs graph*) as a subgraph of a graph G in which each component is regular with each vertex of degree 1 or 2. In other words, the components of sesquivalent subgraphs are either single edges (i.e., regular graphs with each vertex of degree 1) or circuits (i.e., regular graphs with each vertex of degree 2). Then, the coefficients a_n of $P_G(x)$ in Equation 1–2 can be defined by the following equations:

$$a_0 = 1 \tag{1–4a}$$

---

[11]F. Harary, *SIAM Rev.*, **4**, 202, 1962.
[12]R. H. Bruck, *Pac. J. Math.*, **13**, 421, 1963.
[13]J. A. Baker, *J. Math. Phys.*, **7**, 2238, 1966.
[14]M. Fisher, *J. Comb. Theor.*, **1**, 105, 1966.
[15]J. Ponstein, *SIAM J. Appl. Math.*, **14**, 600, 1966.
[16]J. Turner, *SIAM J. Appl. Math.*, **16**, 520, 1968.
[17]D. Ž. Djoković, *Acta Math. Acad. Sci. Hung.*, **21**, 267, 1970.
[18]A. T. Balaban and F. Harary, *J. Chem. Doc.*, **11**, 258, 1971.
[19]H. Sachs, *Publ Math. (Debrecen)*, **11**, 119, 1963.
[20]N. Biggs, *Algebraic Graph Theory*, Cambridge University Press, London, 1974, Chap. 7.

$$a_k = \sum_{s \in S_k} (-1)^{c(s)} 2^{r(s)} \qquad (1 \leq n \leq v) \qquad (1\text{–}4b)$$

In Equation 1–4b, the parameters have the following significance: S_k is the set of all sesquivalent subgraphs of the graph G with k vertices, and therefore the summation goes over all sesquivalent subgraphs with k vertices; $c(s)$ is the number of components of the sesquivalent subgraphs (i.e., the number of disconnected fragments); $r(s)$ is the number of circuits (rings) in the sesquivalent subgraph. When the set S_k is an empty set, there is no sesquivalent subgraph with k vertices, hence $a_k = 0$ in such cases.

To help in understanding the foregoing algorithm, it is instructive to notice the following:

1. The set S_1 (sesquivalent subgraphs with one vertex) consists of loops that correspond to an edge that starts and finishes at the same vertex. Because these are not normally found in the graphs used to represent the bonding topology in chemical systems, the coefficient a_1 in the characteristic polynomial is generally zero; that is, there is no x^{v-1} term in the characteristic polynomial for a graph G with v vertices;
2. The set S_2 (sesquivalent subgraphs with two vertices) consists of the edges of the graph. Therefore the coefficient a_2 in the characteristic polynomial is $-e$ where e is the number of edges in the graph;
3. The set S_3 (sesquivalent subgraphs with three vertices) consists of the triangles of the graph (i.e., circuits with three edges). Therefore, by Equation 1–4b the coefficient a_3 in the characteristic polynomial is $-2t$, where t is the number of triangles in the graph;
4. The sets S_n ($n \geq 4$) are potentially more complicated, since two or more different types of sesquivalent subgraphs are possible when four or more vertices are involved. For example, the set S_4 can contain pairs of disjoint edges [i.e., $c(s) = 2$ and $r(s) = 0$ in equation 1–4b] as well as closed quadrilateral circuits [i.e., $c(s) = 1$ and $r(s) = 1$ in Equation 1–4b]. In the use of the Sachs algorithm, the greatest risk of error arises in the calculation of the a_k ($k \geq 4$) coefficients because of the greater complexity of the corresponding set S_k, which makes it easier to overlook some of the relevant sesquivalent subgraphs.

As a simple illustration of the problem of determining the spectrum of a graph, consider the C_4 (square) graph, which is depicted in the previous section. The adjacency matrix of the C_4 graph is the following 4×4 matrix:

$$\mathbf{A}(\mathrm{C}_4) = \begin{pmatrix} 0 & 1 & 0 & 1 \\ 1 & 0 & 1 & 0 \\ 0 & 1 & 0 & 1 \\ 1 & 0 & 1 & 0 \end{pmatrix} \qquad (1\text{–}5)$$

From this adjacency matrix, the following determinantal equation (1–2) can be obtained :

$$x\mathbf{I} - \mathbf{A} = \begin{pmatrix} x & -1 & 0 & -1 \\ -1 & x & -1 & 0 \\ 0 & -1 & x & -1 \\ -1 & 0 & -1 & x \end{pmatrix} \tag{1-6}$$

$$\Rightarrow \qquad |x\mathbf{I} - \mathbf{A}| = x^4 - 4x^2 = x^2(x^2 - 4) = 0 \Rightarrow x = 2, 0, 0, -2 \tag{1–7}$$

In deriving Equation 1–7, the following sesquivalent subgraphs are used to obtain the coefficients of the characteristic polynomial $P_G(x)$:
a_2: The sesquivalent subgraphs with two vertices of the square are the six edges, each of which consists of one component and no rings—therefore $a_2 = (4)(-1)^1(2)^0 = -4$ which is the coefficient of x^2 in Equation 1–7.
a_4: There are three sesquivalent subgraphs with four vertices of the square, namely two different disjoint pairs of edges and the square itself, which may be depicted as follows:

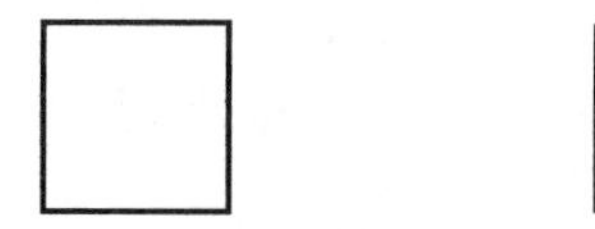

The square contains one component and one ring, whereas the disjoint pairs of edges contain two components and no rings. Therefore:

$$a_4 = (1)(-1)^1(2)^1 + (2)(-1)^2(2)^0 = -2 + 2 = 0 \tag{1–8}$$

so that the coefficient a_4 drops out. The characteristic polynomial thus has no constant term. The spectrum of the square as obtained from Equation 1–7 can be depicted graphically as follows:

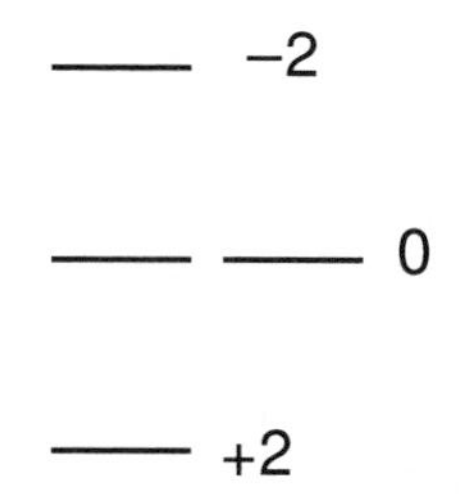

Such a diagram suggests a relationship between graph spectra and energy levels in chemical systems modeled by the graph in question; this relationship will arise frequently later in this book.

The characteristic polynomial (Equation 1–2) of a graph G with v vertices is a polynomial of degree v. In the general case of a graph G with no symmetry, this polynomial may not be factorable. In order to determine the spectrum of the graph and the energy levels of the corresponding chemical system in such a case, it is necessary to solve an equation of degree v.

Fortunately, most chemical systems of interest have some symmetry. The presence of any symmetry in a graph G, even a single twofold symmetry element such as a C_2 rotation axis, an inversion center (i), or a reflection plane (σ), makes the characteristic polynomial factorable into a product of polynomials of smaller degree, such that the sum of the degrees of the factors equals the number of vertices of the original graph. Algorithms have been reported[21,22,23,24,25,26,27,28,29] for the factoring of the characteristic polynomials of graphs using their two- and three-fold symmetry elements. Such symmetry factoring methods use symmetry elements to convert a connected graph G with v vertices into a disconnected graph G^* also with v vertices but with c components $G_1,\ldots,G_c$ such that the spectrum of G^* is the same as that of G. However, whereas the determination of the spectrum of G requires solution of a single equation of degree v, determination of the equivalent spectrum of G^* requires solution of several equations, but none with degrees higher than $u < v$, where u is the number of vertices in the largest component of G^*. Even if G is the usual type of graph with all edges of unit weight in both directions, the corresponding graph G^* arising from symmetry factoring algorithms may have edges of variable and non-unit weights, different edge weights in each direction, and/or vertex loops. However, the complications added by introduction of these features into G^* from G are far less than the simplifications in the expansion of the adjacency matrix into the characteristic polynomial and the reduction of the maximum degree of the equations that must be solved in order to obtain the graph spectrum.

1.4 POLYHEDRAL TOPOLOGY

The concept of a graph leads very naturally to the concept of a polyhedron, which essentially is a way of imbedding a graph into three-dimensional space.

---

[21] E. Heilbronner, *Helv. Chim. Acta*, **37**, 913, 1954.
[22] B. J. McClelland, *J. Chem. Soc. Faraday Trans. 2*, **70**, 1453, 1974.
[23] B. J. McClelland, *J. Mol. Phys.*, **45**, 189, 1982.
[24] B. J. McClelland, *J. Chem. Soc. Faraday Trans. 2*, **78**, 911, 1982.
[25] G. G. Hall, *J. Mol. Phys.*, **33**, 551, 1977.
[26] R. B. King, *Theor. Chim. Acta*, **44**, 223, 1977.
[27] S. S. D'Amato, *Mol. Phys.*, **37**, 1363, 1979.
[28] S. S. D'Amato, *Theor. Chim. Acta*, **53**, 319, 1979.
[29] R. A. Davidson, *Theor. Chim. Acta*, **58**, 193, 1981.

A graph G forming a polyhedron P may be called the *1-skeleton*[30] of the polyhedron; the vertices and edges of G form the vertices and edges of P. The imbedding of G into space to form P leads to the two-dimensional *faces* of P. Of fundamental importance are the following elementary relationships between the numbers and types of vertices (v), edges (e), and faces (f) of polyhedra:[31]

1. Euler's relationship:

$$v - e + f = 2 \qquad (1\text{–}9)$$

This arises from the properties of ordinary three-dimensional space.

2. Relationship between the edges and faces:

$$\sum_{i=3}^{v-1} i f_i = 2e \qquad (1\text{–}10)$$

In Equation 1–10, f_i is the number of faces with i edges (i.e., f_3 is the number of triangular faces, f_4 is the number of quadrilateral faces, etc.). This relationship arises from the fact that each edge of the polyhedron is shared by exactly two faces. Since no face can have fewer edges than the three of a triangle, the following inequality must hold in all cases:

$$3f \leq 2e \qquad (1\text{–}11)$$

3. Relationship between the edges and vertices:

$$\sum_{i=3}^{v-1} i v_i = 2e \qquad (1\text{–}12)$$

In Equation 1–12, v_i is the number of vertices of *degree i* (i.e., having i edges meeting at the vertex). This relationship arises from the fact that each edge of the polyhedron connects exactly two vertices. Since no vertex of a polyhedron can have a degree less than three, the following inequality must hold in all cases:

$$3v \leq 2e \qquad (1\text{–}13)$$

4. Totality of faces:

$$\sum_{i=3}^{v-1} f_i = f \qquad (1\text{–}14)$$

---

[30] B. Grünbaum, *Convex Polytopes*, Interscience, New York, 1967.

[31] R. B. King, *J. Am. Chem. Soc.*, **91**, 7211, 1969.

5. Totality of vertices:

$$\sum_{i=3}^{v-1} v_i = v \qquad (1\text{–}15)$$

Equation 1–14 relates the f_i's to f and Equation 1–15 relates the v_i's to v.

In generating actual polyhedra, the operations of capping and dualization are often important. *Capping* a polyhedron P_1 consists of adding a new vertex above the center of one of its faces F_1 followed by adding edges to connect the new vertex with each vertex of F_1. This capping process gives a new polyhedron P_2 having one more vertex than P_1. If a triangular face is capped, the following relationships will be satisfied where the subscripts 1 and 2 refer to P_1 and P_2, respectively: $v_2 = v_1 + 1$; $e_2 = e_1 + 3$; $f_2 = f_1 + 2$. Such a capping of a triangular face is found in the capping of an octahedron to form a capped octahedron, i.e.:

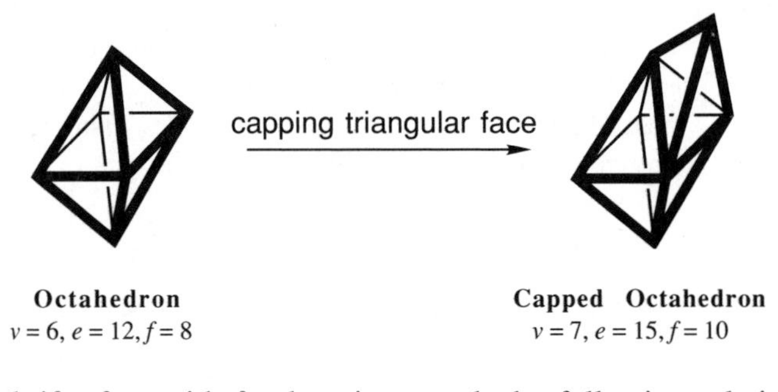

Octahedron
$v = 6, e = 12, f = 8$

Capped Octahedron
$v = 7, e = 15, f = 10$

In general, if a face with f_k edges is capped, the following relationships will be satisfied: $v_2 = v_1 + 1$; $e_2 = e_1 + f_k$; $f_2 = f_1 + f_k - 1$. An example of such a capping process converts a pentagonal pyramid into a pentagonal bipyramid, i.e.,

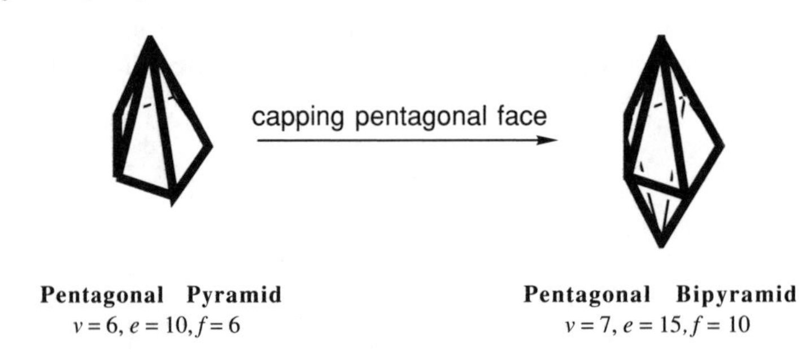

Pentagonal Pyramid
$v = 6, e = 10, f = 6$

Pentagonal Bipyramid
$v = 7, e = 15, f = 10$

A given polyhedron P can be converted into its dual P^* by locating the centers of the faces of P^* at the vertices of P and the vertices of P^* above the centers of the faces of P. Two vertices in the dual P^* are connected by an edge when the corresponding faces in P share an edge. An example of the process of dualization is the conversion of a regular octahedron to a cube, i.e.,

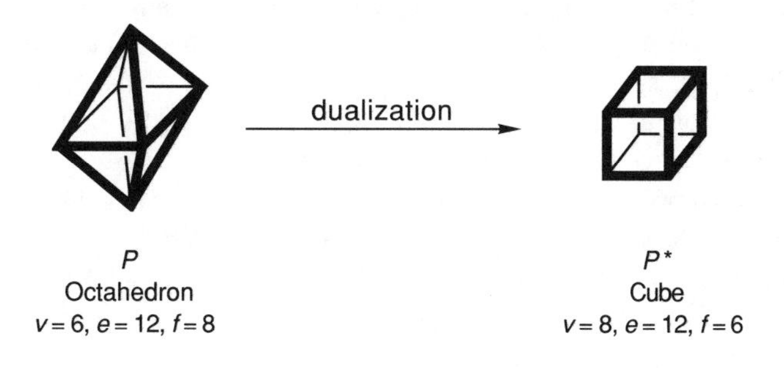

The process of dualization has the following properties:

1. The numbers of vertices and edges in a pair of dual polyhedra P and P^* satisfy the relationships $v^* = f$, $e^* = e$, $f^* = v$.
2. Dual polyhedra have the same symmetry elements and thus belong to the same symmetry point group.
3. Dualization of the dual of a polyhedron leads to the original polyhedron.
4. The degrees of the vertices of a polyhedron correspond to the number of edges in the corresponding face polygons in its dual.

Two polyhedra, $\mathcal{P}_1$ and $\mathcal{P}_2$, may be considered to be *combinatorially equivalent*[30] whenever there are three one-to-one mappings $\mathcal{V}$, $\mathcal{E}$, and $\mathcal{F}$ from the vertex, edge, and face sets of $\mathcal{P}_1$ to the corresponding sets of $\mathcal{P}_2$, such that incidence relations are conserved. Thus, if a vertex, edge or face α of $\mathcal{P}_1$ is incident to or touches upon a vertex, edge, or face β of $\mathcal{P}_1$, then the images of α and β under $\mathcal{V}$, $\mathcal{E}$, or $\mathcal{F}$ are incident in $\mathcal{P}_2$.[32]

In this chapter and elsewhere in this book, polyhedra are depicted as two-dimensional "perspective" drawings as aids to help visualize the actual three-dimensional structures. For more complicated polyhedra, these two-dimensional perspective drawings begin to have limitations, since both their drawing requires skill and reconstruction of their original three-dimensional picture requires more imagination. These difficulties can be minimized by the use of Schlegel diagrams[30,33] rather than conventional perspective drawings to depict three-dimensional polyhedra in two dimensions. Schlegel diagrams are well known to mathematicians studying polyhedra and higher dimension polytopes but are relatively unfamiliar to chemists.

In order to obtain a Schlegel diagram of a polyhedron $\mathcal{P}$, select any face of $\mathcal{P}$ as the *base face*, $\mathcal{F}_0$. The plane containing the base face $\mathcal{F}_0$ separates three-dimensional space into two half-spaces, one of which contains the entire

---

[32] X. Liu, D. J. Klein, T. G. Schmalz, and W. A. Seitz, *J. Comput. Chem.*, **12**, 1252, 1991.
[33] V. Schlegel, *Nova Acta Leop. Carol.*, **44**, 343, 1883.

volume of $\mathcal{P}$. Select a point x_0 in the other half-space. Draw a straight line from x_0 to each of the vertices of $\mathcal{P}$. Each such line will intersect the plane of $\mathcal{F}_0$ at a point representing the corresponding vertex. Connect a pair of vertex projections onto the plane of $\mathcal{F}_0$ with straight lines if and only if the corresponding vertices of $\mathcal{P}$ have an edge between them. This process leads to a projection of the three-dimensional plane of the face $\mathcal{F}_0$; this projection is called the *Schlegel diagram* of the polyhedron $\mathcal{P}$.

Any given polyhedron can have as many different Schlegel diagrams as it has different faces. The procedure for drawing the Schlegel diagram of the square pyramid using the square face as the base face $\mathcal{F}_0$ is illustrated below.

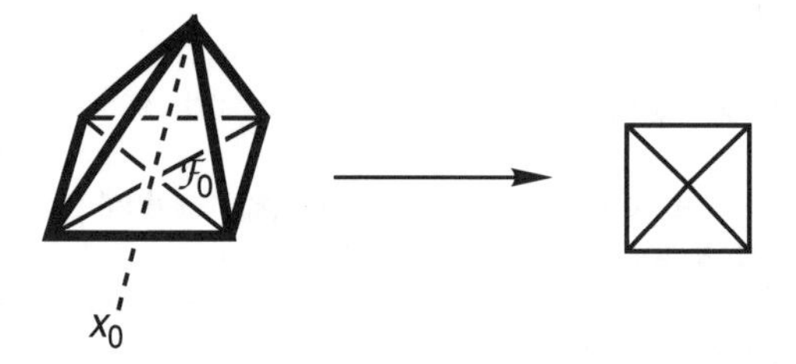

The following features of Schlegel diagrams are of interest:

1. The location of the point x_0 can always be chosen so that the edges in the Schlegel diagram can be drawn as *non-intersecting* straight lines. This is one of the big advantages of Schlegel diagrams over conventional perspective drawings.
2. Schlegel diagrams depict the topological but not the metric features of polyhedra. Thus, the vertex neighborhood relationships depicted by edges are preserved. However, edge lengths and angles are distorted. Since many important chemical relationships are topological rather than metric, this distortion is not necessarily serious.
3. Schlegel diagrams may not preserve all symmetry elements of the original polyhedron because of the metric distortion. The preservation of symmetry elements in Schlegel diagrams is maximized if a unique face of the polyhedron is selected as the base face.

The regular octahedron gives the following Schlegel diagram:

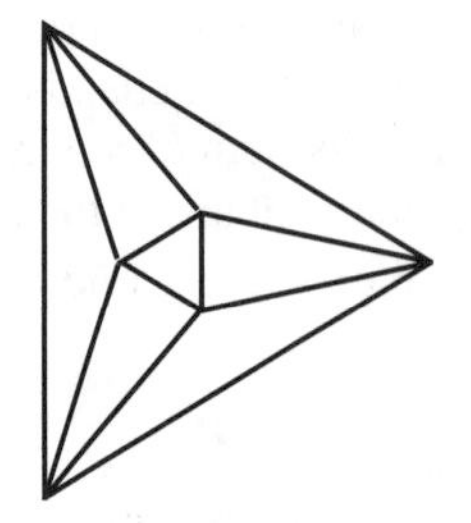

The big triangle corresponds to the base face and contains all of the other faces in the Schlegel diagram. The small triangle in the center is the projection of the face opposite to the base face. The other triangles correspond to the remaining six faces of the octahedron. This Schlegel diagram of an octahedron shows how Schlegel diagrams fail to preserve metric and symmetry properties since, in a regular octahedron, all of the eight triangular faces are congruent and all of the twelve edges have the same length.

These Schlegel diagrams of polyhedra are graphs whose vertices and edges correspond to those of the polyhedron. With the standard convention that polyhedra are homeomorphic to the sphere,[34] the corresponding Schlegel diagram can be imbedded in the surface of a sphere. Punching a hole through the surface of the sphere without touching any components of the Schlegel diagram followed by topologically deforming the remaining surface to a disk shows that the Schlegel diagram is a *planar* graph (Section 1.2). This process is reversible, and the Schlegel diagram obtained is unique,[35] although it may appear somewhat different depending upon the face in which the hole in punched.

The problem of classification and enumeration of polyhedra is a complicated one. Thus, there appear to be no formulas, direct or recursive, for which the number of combinatorially (topologically) distinct polyhedra having a given number of vertices, edges, faces, or any combination of these elements can be calculated.[36,37] Duijvestijn and Federico have enumerated by computer the polyhedra having up to 22 edges according to their numbers of vertices, edges, and faces and their symmetry groups and present a summary of their methods, results, and literature references to previous work.[38] Their work shows that there are 1, 2, 7, 34, 257, 2606, and 32,300 topologically distinct polyhedra, having 4, 5, 6, 7, 8, 9, and 10 faces or vertices, respectively. Tabulations are available for all 301 (= 1 + 2 + 7 + 34 + 257) topologically distinct polyhedra, having eight or fewer faces[39] or eight or fewer vertices.[40] These two tabulations are essentially equivalent by the dualization relationship discussed above. Some of the properties of the polyhedra having six or fewer vertices are listed in Table 1–2.

---

[34]M. J. Mansfield, *Introduction to Topology*, D. Van Nostrand, Princeton, New Jersey, 1963, 40.

[35]H. Whitney, *Am. J. Math.*, **54**, 150, 1932.

[36]F. Harary and E. M. Palmer, *Graphical Enumeration*, Academic Press, New York, 1973, 224.

[37]W. T. Tutte, *J. Combin. Theor. Ser. B,* **28**, 105,, 1980.

38A. J. W. Duijvestijn and P. J. Federico, *Math. Comput.*, **37**, 523, 1981.

[39]P. J. Federico, *Geom. Ded.*, **3**, 469, 1975.

[40]D. Britton and J. D. Dunitz, *Acta Cryst.*, **A29**, 362, 1973.

TABLE 1–2
Some Properties of All Polyhedra with Six or Fewer Vertices

					Vertices			Faces		
Polyhedron Name	v	e	f	Symmetry	v_3	v_4	v_5	f_3	f_4	f_5
Tetrahedron	4	6	4	T_d	4	0	0	4	0	0
Trigonal bipyramid	5	9	6	D_{3h}	2	3	0	6	0	0
Square pyramid	5	8	5	C_{4v}	4	1	0	4	1	0
Octahedron	6	12	8	O_h	0	6	0	8	0	0
Bicapped tetrahedron	6	12	8	C_{2v}	2	2	2	8	0	0
	6	11	7	C_{2v}	2	4	0	6	1	0
	6	11	7	C_s	3	2	1	6	1	0
	6	10	6	C_2	4	2	0	4	2	0
Pentagonal pyramid	6	10	6	C_{5v}	5	0	1	5	0	1
Trigonal prism	6	9	5	D_{3h}	6	0	0	2	3	0

Chapter 2

SYMMETRY AND GROUP THEORY

2.1 SYMMETRY OPERATIONS

In many of the applications of topology and graph theory to problems in inorganic chemistry, the concept of *symmetry* arises naturally. This chapter summarizes some of the fundamental ideas of symmetry and the use of group theory to study various aspects of symmetry. For background information or further details, the reader is referred to standard texts on chemical applications of group theory.[1]

A *symmetry operation* is a movement of an object such that, after completion of the movement, every point of the body coincides with an equivalent point or the same point of the object in its original orientation. The position and orientation of an object before and after carrying out a symmetry operation are indistinguishable. Thus a symmetry operation takes an object into an equivalent configuration.

The symmetry operations for objects in ordinary three-dimensional space can be classified into four fundamental types, each of which is defined by a *symmetry element* around which the symmetry operation takes place. The four fundamental types of symmetry operations and their corresponding symmetry elements are listed in Table 2–1.

TABLE 2–1
The Four Fundamental Types of Symmetry Operations

Symmetry Operation	Designation	Corresponding Symmetry Element	Dimensions
Identity (no change)	E	The entire object	3
Reflection	σ	Reflection plane	2
Rotation	C_n	Rotation axis	1
Improper rotation	S_n	Improper rotation axis (= point of intersection of a proper rotation axis and a perpendicular reflection plane)	0

[1] F. A. Cotton, *Chemical Applications of Group Theory*, John Wiley, & Sons New York, 1971.

The identity operation, designated as E, leaves the object unchanged. Although this operation may seem trivial, it is mathematically necessary in order to convey the mathematical properties of a group on the set of all of the symmetry operations applicable to a given object. This point will be clarified later in this chapter. The reflection operation, designated as σ, involves reflection of the object through a plane, known as a *reflection plane*. For example, in a reflection through the xy-plane (conveniently designated as σ_{xy}) the coordinates of a point (x, y, z) change to $(x, y, -z)$; a reflection operation thus can result in the change of only a single coordinate. A rotation operation, designated as C_n, consists of a $360°/n$ rotation around a line, known as a *rotation axis*. For example, a C_2 rotation around the z-axis changes the coordinates of a point (x, y, z) to $(-x, -y, z)$; a rotation operation thus can result in a change of only two coordinates. An improper rotation, designated as S_n, consists of a $360°/n$ rotation around a line followed by a reflection in a plane perpendicular to the rotation axis. An S_2 operation is called an *inversion* and is designated by i; the intersection of the C_2-axis and the perpendicular reflection plane is called an *inversion center*. Inversion through the origin changes the coordinates of a point (x, y, z) to $(-x, -y, -z)$; thus an S_n operation must change the signs of all three coordinates. An S_1 improper rotation in which the C_1 proper rotation component is equivalent to the identity E corresponds to a reflection operation σ. Thus the reflection operation σ is a special type of improper rotation, namely S_1.

2.2 SYMMETRY POINT GROUPS

Consider the set of symmetry operations in ordinary three-dimensional space describing the symmetry of an actual object such as a molecule. Such a set of symmetry operations satisfies the properties of a *group* in the mathematical sense and is therefore called a *symmetry point group*. In most cases such a symmetry point group contains a finite number of operations and is therefore a *finite group*. The properties of such groups will be considered in this section.

A set of operations forming a mathematical group must satisfy the following four conditions:

1. **The product of any two operations in the group and the square of each operation must be an operation of the group.** In order to apply this condition, the concept of a product of operations must be defined. In this connection a *product* of two symmetry operations is obtained by applying them successively. A *square* of a symmetry operation is obtained by applying the same operation twice. This definition can be extended to higher powers of symmetry operations. Forming a product of two group operations is called *multiplication* by analogy to arithmetic. The multiplication of two operations in a group is said to be *commutative* if the order of

multiplication is immaterial, i.e., if $AB = BA$. In such a case, A is said to *commute* with B. The multiplication of two symmetry operations is *not* necessarily commutative.

2. **One operation in the group must commute with all others and leave them unchanged.** This operation is conventionally called the *identity operation.* In this case of symmetry point groups, the identity operation consists of "doing nothing" and is conventionally designated as E (Table 2–1). This is why the "trivial" identity operation E must be considered when treating symmetry point groups. This condition may be concisely stated as $EX = XE = X$.
3. **The associative law of multiplication must hold.** This condition may be expressed concisely as $A(BC) = (AB)C$, i.e., the result must be the same if C is multiplied by B to give BC followed by multiplication of BC by A to give $A(BC)$ or if B is multiplied by A to give AB followed by multiplication of AB by C to give $(AB)C$.
4. **Every operation must have a reciprocal, which is also an operation in the group.** The operation Z is the *reciprocal* of the operation A if $AZ = ZA = E$. The reciprocal of an operation A is frequently designated by A^{-1}. Note that multiplication of an operation by its reciprocal is always commutative.

These defining characteristics of a group have been summarized concisely[2] by defining a group as "...a mathematical system consisting of elements with *inverses* which can be combined in some operation without going outside the system."

The operations in a group can be characterized by their periods. In this context the *period* of an operation is the minimum number of times it must be multiplied by itself before the identity operation E is obtained. In the case of symmetry operations (Table 2–1) the period of the identity operation E is, of course, 1; the periods of the reflections σ are always 2; the periods of the proper rotations C_n are n; the periods of the *even* improper rotations S_{2n} are $2n$; and the periods of the *odd* improper rotations S_{2n+1} are $4n + 2$.

The number of operations in a group is sometimes called the *order* of the group. Within a given group, it may be possible to select various smaller sets of operations, each set including the identity element E, which are themselves groups. Such smaller sets are called *subgroups.* A subgroup of a group G is thus defined as a subset H of the group G, which is itself a group under the multiplication of G. The fact that H is a subgroup of G may be written $H \subset G$. The order of a subgroup must be an integral factor of the order of the group. Thus if H is a subgroup of G and $|H|$ and $|G|$ are the orders of H and G, respectively, then the quotient $|G|/|H|$ must be an integer; this quotient is called the *index* of the subgroup H in G.

---

[2]A. W. Bell and T. J. Fletcher, *Symmetry Groups*, Associated Teachers of Mathematics, 1964.

Let A and X be two operations in a group. Then $X^{-1}AX = B$ will be equal to some operation in the group. The operation B is called the *similarity transform* of A by X, and A and B may be said to be *conjugate*. Conjugate operations have the following properties:

1. **Every operation is conjugate with itself.** Thus for any particular operation A there must be at least one operation X such that $A = X^{-1} AX$.
2. **If A is conjugate with B, then B is conjugate with A.** Thus if $A = X^{-1} BX$, then there must be some element, Y, in the group such that $B = Y^{-1} AY$.
3. **If A is conjugate with B and C, then B and C are conjugate with each other.**

A complete set of operations of a group which are conjugate to one another is called a *class* (or more specifically a *conjugacy class*) of the group. The number of operations in a conjugacy class is called its *order*; the orders of all conjugacy classes must be integral factors of the order of the group.

A group in which every operation commutes with every other operation is called a *commutative* group, or an *Abelian* group after the famous Norwegian mathematician, Abel (1802–1829). In an Abelian group, every operation is in a conjugacy class by itself; i.e., all conjugacy classes are of order one. A *normal subgroup* N of G, written $N \lhd G$, is a subgroup which consists only of *entire* conjugacy classes of G.[3] A *normal chain* of a group G is a sequence of normal subgroups $C_1 \lhd N_{a_1} \lhd N_{a_2} \lhd N_{a_3} \lhd \cdots \lhd N_{a_s} \lhd G$, in which s is the number of normal subgroups (besides C_1 and G) in the normal chain. A *simple* group has no normal subgroups other than the identity group C_1. Simple groups are particularly important in the theory of finite groups.[4] If a normal chain starts with the identity group C_1 and leads to G and if all of the quotient groups $N_{a_1} / C_1 = C_{a_1}, N_{a_2} / N_{a_1} = C_{a_2} \ldots, G / N_{a_s} = C_{a_{s+1}}$ are cyclic, then G is a *composite* or *soluble* group. A soluble group can be expressed as a *direct product* of the factor groups $C_{a_1} \times C_{a_2} \times \ldots \times C_{a_{s+1}}$.

Any *complete* set of symmetry operations for an object satisfies the above four conditions for a mathematical group. Such a group is called a *symmetry point group*. Such symmetry point groups have a standard designation, called the *Schoenflies symbol*, based on either a conspicuous operation in the point group (such as the highest order rotation axis) or a common case of occurrence of the symmetry point group (such as describing the symmetry of a well-known polyhedron like the tetrahedron, octahedron, or icosahedron).

---

[3]J. K. G. Watson, *Mol. Phys.*, **21**, 577, 1971.

[4]D. Gorenstein, *Finite Groups*, Harper & Row, New York, 1968.

There is a systematic way to classify static objects in three-dimensional space by their symmetry point groups based on the following sequence of questions:[1]

1. **Is the object linear (i.e., only "one-dimensional")?** Linear objects are the only objects having infinite rather than finite symmetry point groups, since the linear axis corresponds to an infinite order rotation axis, namely C_∞. If there is a reflection plane perpendicular to the infinite order rotation axis (dividing the object into two equivalent halves), then the symmetry point group is D_∞; if not, the symmetry point group is C_∞.
2. **Does the object have multiple "higher-order" rotation axes (i.e., $C_{>2}$ axes)?** Multiple axes in this question refer to non-collinear $C_{>2}$ axes. Such non-linear objects have the symmetries of well-known regular polyhedra and are generally readily recognizable in containing the regular polyhedra in some manner. Such groups are sometimes called the *polyhedral* point groups. Thus the tetrahedral groups T, T_h, and T_d, with 12, 24, and 24 operations, respectively, have four non-collinear C_3 axes and are distinguished by the presence or absence of horizontal (σ_h) or diagonal (σ_d) reflection planes. The octahedral groups O and O_h, with 24 and 48 operations, respectively, have not only four non-collinear C_3 axes but also three non-collinear C_4 axes and are distinguished by the presence or absence of reflection planes. The icosahedral groups I and I_h, with 60 and 120 operations, respectively, have ten non-collinear C_3 axes and six non-collinear C_5 axes and are also distinguished by the presence or absence of reflection planes.
3. **If the object does not belong to either a linear group or a polyhedral group, then does it have proper or improper axes of rotation (i.e. C_n or S_n)?** If no axes of either type are found, the group is either C_s, C_i, or C_1 with 2, 2, and 1 operations, respectively, depending upon whether the object has a plane of symmetry (σ), and inversion center (i), or neither. The C_1 designation corresponds to an object with no symmetry at all and, therefore, to a symmetry point group containing only one element, namely the identity element E.
4. **Does the object have an *even*-order improper rotation axis S_{2n} but no planes of symmetry or any proper rotation axis other than one collinear with the improper rotation axis?** The presence of an even-order improper rotation axis S_{2n} without any non-collinear proper rotation axes or any reflection planes indicates the symmetry point group S_{2n} with $2n$ operations.
5. **If the object does not belong to the linear point groups, the polyhedral point groups, or the point groups C_s, C_n,**

C_1, or S_{2n}, then look for the highest order rotation axis. Call the highest order rotation axis C_n.

6. **Are there n C_2 axes lying in a plane perpendicular to the C_n axis?** If there are n C_2 axes lying in a plane perpendicular to the C_n axis, then the object belongs to one of the symmetry point groups D_n, D_{nh}, or D_{nd} with $2n$, $4n$, and $4n$ operations, respectively, depending on whether there are no planes of symmetry, a horizontal plane of symmetry (σ_h), or n vertical planes of symmetry (σ_v), respectively. If there are no C_2 axes lying in a plane perpendicular to the C_n axis, then the object belongs to one of the symmetry point groups C_n, C_{nv}, C_{nh}, with n, $2n$, and $2n$ operations, respectively, depending upon whether there are no planes of symmetry, n vertical planes of symmetry (σ_v), or a horizontal plane of symmetry (σ_h), respectively. If there are only C_2 axes, then a unique C_2 axis is chosen as the "reference axis", if there is any ambiguity as to which C_2 axis to choose.

Some of these ideas are illustrated using the trigonal bipyramid depicted below with the locations of the C_3 and three C_2 axes, indicated by a solid arrow and three dotted arrows, respectively.

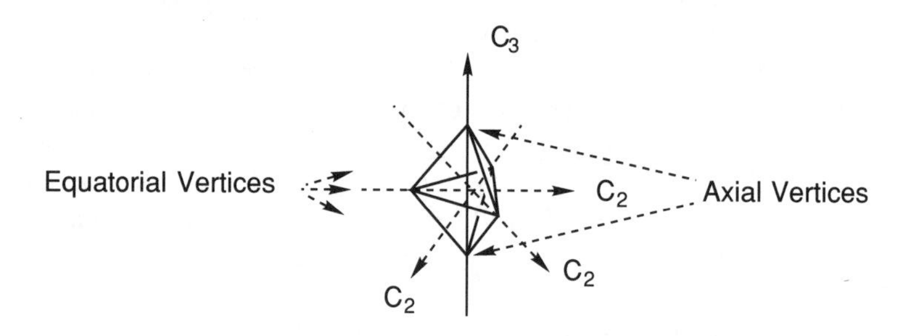

The symmetry point group of the trigonal bipyramid is D_{3h} of order 12, with the conjugacy classes E, $2C_3$, $3C_2$, σ_h, $2S_3$, $3\sigma_v$. Note the following concerning these conjugacy classes:

1. The identity operation E is in a class by itself.
2. The two operations in the C_3 class correspond to 120° (C_3) and 240° (C_3^2) rotations around the single C_3 axis. Note that the C_3^2 rotation is the inverse of the C_3 rotation, i.e., $C_3^{-1} = C_3^2$.
3. The C_2 rotations around each of the three C_2 axes form the class $3C_2$ of order 3.
4. Reflection in the plane of the equatorial vertices (σ_h) is in a class by itself.
5. The two operations in the S_3 class correspond to 120° and 240° rotations around the C_3 axis, followed by reflection in the σ_h plane. Thus $S_3 = (\sigma_h)(C_3)$ and $S_3^2 = (\sigma_h)(C_3^2)$ are two examples of multiplication operations in the D_{3h} point group.

6. The three types of reflections in the planes containing both axial vertices and one of the three equatorial vertices (σ_v) form the class $3\sigma_v$ of order 3.

2.3 GROUP REPRESENTATIONS

A *representation* of a group is defined as a set of matrices, each corresponding to a single operation in the group, that can be combined among themselves by matrix multiplication in a manner analogous to the way the group operations combine by the group multiplication operation. The matrices will necessarily be square $d \times d$ matrices with d being the dimension of the group representation. One-dimensional group representations thus correspond to scalar quantities, which, in order for all of the group properties to be satisfied, can only be 1, –1, and the various roots of unity of the type $e^{2\pi i/n}$ including $i(=\sqrt{-1})$ and $-i$.

Since square matrices with any integral number of dimensions are possible, the number of possible (matrix) representations of any group is infinite. However, a small set of matrix representations have properties of particular significance; these are called the *irreducible representations* of the group (sometimes abbreviated as *irred reps*).

Consider a set of matrices **E**, **A**, **B**, **C**,... which form a representation of a given group. Any similarity transformation can be applied to each matrix in this set to give a new set of matrices, i.e., $\mathbf{E}' = \mathbf{Q}^{-1}\mathbf{E}\mathbf{Q}$, $\mathbf{A}' = \mathbf{Q}^{-1}\mathbf{A}\mathbf{Q}$, $\mathbf{B}' = \mathbf{Q}^{-1}\mathbf{B}\mathbf{Q}$, $\mathbf{C}' = \mathbf{Q}^{-1}\mathbf{C}\mathbf{Q}$,..., which is a new matrix representation of the same group. In some cases application of a similarity transformation to a representation will give a *block-factored* matrix, e.g., equation 2–1 for a matrix **A** blocked into five smaller matrices $\mathbf{A}_1'$, $\mathbf{A}_2'$, $\mathbf{A}_3'$, $\mathbf{A}_4'$, and $\mathbf{A}_5'$ where **0** is a matrix with all zero entries:

$$\mathbf{A}' = \mathbf{Q}^{-1}\mathbf{A}\mathbf{Q} = \begin{pmatrix} \boxed{\mathbf{A}_1'} & \mathbf{0} & \mathbf{0} & \mathbf{0} & \mathbf{0} \\ \mathbf{0} & \boxed{\mathbf{A}_2'} & \mathbf{0} & \mathbf{0} & \mathbf{0} \\ \mathbf{0} & \mathbf{0} & \boxed{\mathbf{A}_3'} & \mathbf{0} & \mathbf{0} \\ \mathbf{0} & \mathbf{0} & \mathbf{0} & \boxed{\mathbf{A}_4'} & \mathbf{0} \\ \mathbf{0} & \mathbf{0} & \mathbf{0} & \mathbf{0} & \boxed{\mathbf{A}_5'} \end{pmatrix} \qquad (2\text{–}1)$$

If each of the matrices in the representation is blocked out in the same way by the similarity transformation, then corresponding blocks of each matrix can be multiplied together separately. The set of matrices **E**, **A**, **B**, **C**,... is called a *reducible* representation because it is possible to transform each matrix into a new matrix so that all of the new matrices can be decomposed in the same way to two or more representations of a smaller dimension. If it is not possible to find a similarity transformation which can reduce all of the matrices of a given representation in the same manner, then the representation is said to be *irreducible*.

In practice it is relatively inconvenient to handle the matrices corresponding to a group representation. In general, it is sufficient to use the characters of these matrices; the *character* of a matrix is simply the scalar quantity corresponding to the sum of the diagonal elements of the matrix. The use of the character of a matrix is feasible because conjugate matrices have identical characters.

Irreducible representations and their characters of a group have the following properties of interest:

1. **The sum of the squares of the dimensions of the irreducible representations of a group is equal to the order of the group.**
2. **The sum of the squares of the characters in any irreducible representation of a group is equal to the order of the group.**
3. **The vectors whose components are the characters of the various group operations of two different irreducible representations are orthogonal.** In this connection, note that the characters of the matrices forming an irreducible representation of a group G with $|G|$ operations can be described by a $|G|$-dimensional vector.
4. **The characters of all matrices belonging to operations in the same conjugacy class of a given representation are identical.**
5. **The number of irreducible representations of a group is equal to the number of classes of the group.** If r = the number of irreducible representations of a group = the number of its classes, then the characters of the irreducible representations of a group can be listed as an $r \times r$ square matrix in a so-called *character table*. Character tables of the symmetry point groups are generally listed in the back of texts on chemical group theory.

There is a standard set of symbols known as the *Mulliken symbols*, that are used to designate the irreducible representations of symmetry point groups. The Mulliken symbols conform to the following rules:

1. All one-dimensional representations are designated by either A or B; two-dimensional representations are designated by E; three-dimensional representations are designated by T or less frequently F; four-dimensional representations are designated by G; and five-dimensional representations are designated by H.
2. One-dimensional representations which are symmetric with respect to rotation by $2\pi/n$ about the principal C_n axis are designated by A

whereas those which are antisymmetric with respect to rotation by $2\pi/n$ about the principal C_n axis are designated by B.

3. The subscripts 1 and 2 attached to A's and B's designate those which are symmetric or antisymmetric, respectively, with respect to a C_2 axis perpendicular to the principal C_n axis or, in the absence of such a C_2 axis, to a vertical plane of symmetry (σ_v).
4. Primes (′) and double primes (″) are attached to all letters to indicate irreducible representations which are symmetric and antisymmetric, respectively, with respect to the horizontal plane of symmetry (σ_h).

2.4 SYMMETRY POINT GROUPS AS DIRECT PRODUCT GROUPS

Consider the two groups G with m operations $E, g_2,\ldots,g_m$ and H with n operations $E, h_2,\ldots,h_n$, in which the operations of G and H are independent except for the identity. The direct product $G \times H$ contains mn paired operations of the type $EE, g_2E,\ldots,g_mE, Eh_2, g_2h_2,\ldots,g_mh_2, Eh_3,\ldots,g_{m-1}h_n, g_mh_n$ where EE is the identity of $G \times H$ and where, because of the independence of the operations of G and H, the order of the paired operations in $G \times H$ is immaterial.[5]

The direct product $G \times H$ has the following properties:

1. If G has the r conjugacy classes $K_1 = E, K_2,\ldots, K_r$ and H has the s conjugacy classes $L_1 = E, L_2,\ldots,L_s$, then the direct product $G \times H$ has the rs conjugacy classes $K_1L_1 = E, K_2L_1,\ldots,K_rL_1,K_1L_2, K_2L_2,\ldots,K_rL_2, K_1L_3,\ldots,K_{r-1}L_s, K_rL_s$. The irreducible representations and their characters have a similar product structure.
2. The groups G and H are both normal subgroups of their direct product $G \times H$.

The direct product may also be regarded as a special case of the semidirect product $G \wedge H$, in which only the first of the two groups (namely G) needs to be a normal subgroup of the product. Furthermore, the conjugacy classes, irreducible representations, and characters of a semidirect product do not have simple relationships to those of the factors in contrast to a direct product. The full definition of a semidirect product is considerably more complicated than that of a direct product and will not be presented in this book.

All non-trivial symmetry point groups except for C_s and C_i contain one or more proper rotations C_n ($n \geq 2$) in addition to the identity. In addition, the point groups other than C_n, D_n, T, O, and I contain one or more improper rotations S_n, where S_1 is a symmetry plane σ and S_2 is the inversion center i.

[5] L. Jansen and M. Boon, *Theory of Finite Groups. Applications in Physics*, North Holland, Amsterdam, 1967.

All point groups containing improper rotations S_n are semidirect products of the type $R \wedge C_s'$ where C_s' is either C_s (i.e., $E + \sigma$) or C_i (i.e., $E + i$) and R is a group consisting of only the identity and proper rotations. Furthermore, R is a normal subgroup of the semidirect product $R \wedge C_s'$. For convenience, the non-identity operation (namely σ or i) in the factor C_s' can be called a *primary involution* and designated as S'. Note that some point groups can have more than one primary involution. Thus the point group C_{2v} has two different σ_v primary involutions.

Some point groups are *direct* products of the type $R \times C_s'$ in which both R and C_s' are normal subgroups. These are listed in Table 2–2. Because of the direct product structure, the primary involution in such a group is in a class by itself. The character tables of these direct product point groups are $2r \times 2r$ matrices of the following type in which r is the number of classes in R and $\mathbf{X}$ is an $r \times r$ matrix corresponding to the character table of R:

$$\begin{pmatrix} \mathbf{X} & \mathbf{X} \\ \mathbf{X} & -\mathbf{X} \end{pmatrix} \tag{2–2}$$

In the character table (2–2), half of the characters for the primary involution (σ_h or i) of $R \times C_s'$ are equal to the corresponding characters of the identity. The corresponding irreducible representations may be called the *even* or symmetrical irreducible representations, since if the primary involution is an inversion, these irreducible representations are normally designated in character tables with a "g" for "gerade". The remaining half of the characters in (2–2) for the primary involution are the negatives of the corresponding characters of the identity. The corresponding irreducible representations may be called the *odd* or antisymmetrical irreducible representations, since if the primary involution is an inversion, these irreducible representations are normally designated in character tables with a "u" for "ungerade". A conclusion from these observations is that a reducible representation with zero character for the primary involution must be the sum of an equal number of even and odd irreducible representations. More generally let d^+ and d^- be the sums of the dimensions of the even and odd reducible representations, respectively, forming the reducible representation having a character $\chi(S')$ for the primary involution. Then

$$\chi(S') = d^+ - d^- \tag{2–3}$$

Equation (2–3) describes some constraints on the characters for symmetry point groups expressible as direct product groups, which will be used in the next chapter for studying the properties of forbidden polyhedra.

TABLE 2–2
Direct Product Structures of Symmetry Point Groups Having Reflection Planes and/or Inversion Centers

$$C_{2v} = C_2 \times C_s$$

$$C_{2nh} = C_{2n} \times C_i = C_{2n} \times C_s$$

$$C_{(2n+1)h} = C_{2n+1} \times C_s$$

$$D_{2nh} = D_{2n} \times C_i = D_{2n} \times C_s$$

$$D_{(2n+1)h} = D_{2n+1} \times C_s$$

$$D_{(2n+1)d} = D_{2n+1} \times C_i$$

$$S_{4n+2} = C_{2n+1} \times C_i$$

$$T_h = T \times C_i$$

$$O_h = O \times C_i$$

$$I_h = I \times C_i$$

2.5 PERMUTATION GROUPS

Consider a set X of n objects. The set of permutations of these objects (including the identity "permutation") has the structure of a group and is called a *permutation group* of *degree n*.[6] Let G be a permutation group acting on the set X . Let g be any operation in G and x be any object in set X. The subset of X obtained by the action of all operations in G on x is called the *orbit* of x. The operations in G leaving x fixed is called the *stabilizer* of x; it is a subgroup of G and may be abbreviated as g_x. A *transitive* permutation group has only one orbit containing all objects of the set X. Sites permuted by a transitive permutation group are thus equivalent. Transitive permutation groups represent permutation groups of the "highest symmetry" and thus play a special role in permutation group theory.

The maximum number of distinct permutations of n objects is $n!$. The corresponding group is called the *symmetric* group of degree n and is traditionally designated as S_n. However, this designation can easily be confused with the designation S_n used by chemists for an improper rotation, so that the alternative designation of P_n for the symmetric group on n objects causes less confusion in chemical contexts. The symmetric group P_n is

[6]N. L. Biggs, *Finite Groups of Automorphisms*, Cambridge University Press, London, 1971.

obviously the highest symmetry permutation group of degree n. All permutation groups of degree n must be a subgroup of the corresponding symmetric group P_n.

Let us now consider the permutation group structure of the permutations of ligands attached to a polyhedral skeleton, such as in a metal complex of the stoichiometry ML_n.[7] A permutation $\boldsymbol{P}_n$ of n objects can be described by a $2 \times n$ matrix of the general type

$$\boldsymbol{P}_n = \begin{pmatrix} 1 & 2 & 3 & \dots & n \\ p_1 & p_2 & p_3 & \dots & p_n \end{pmatrix} \tag{2–4}$$

In the example of interest, the top row represents polyhedral vertex labels, and the bottom row represents ligand labels. The numbers $p_1, p_2, p_3,\dots, p_n$ can be taken to run through the integers 1, 2, 3,...,n in some sequence. For a given n there are $n!$ possible different $\boldsymbol{P}_n$ matrices. The matrix $\boldsymbol{P}_n^0$, in which the bottom row $p_1, p_2, p_3,\dots, p_n$ has the integers in the natural order 1, 2, 3,...,n (i.e., the bottom row of $\boldsymbol{P}_n^0$ is identical to the top row), can be taken to represent a reference configuration corresponding to the identity operation in the corresponding permutation group.

Permutations can be classified as *odd* or *even* permutations based on how many *pairs* of numbers in the bottom row of the matrix $\boldsymbol{P}_n$ are out of their natural order. Alternatively, if the interchange of a single pair of numbers is called a *transposition*, the parity of a permutation corresponds to the parity of the number of transpositions. Thus, a permutation which is obtained by an odd number of transpositions from the reference configuration is called an *odd* permutation, and a permutation which is obtained by an even number of transpositions from the reference configuration is called an *even* permutation. The identity permutation corresponding to the reference configuration has zero transpositions and is therefore an even permutation by this definition.

A group can be defined relating the $\boldsymbol{P}_n$ matrices for a given n. First redefine the rows of $\boldsymbol{P}_n$ so that the top row represents the reference configuration $\boldsymbol{P}_n^0$ and the bottom row represents the ligand labels in any of the $n!$ possible permutations of the n ligands. These permutations form a group of order $n!$, with the permutation leaving the reference isomer unchanged (i.e., that represented by $\boldsymbol{P}_n^0$ as so redefined), corresponding to the identity operation, E. This permutation group is the symmetric group as defined above.

Now consider the nature of the operations in a symmetric permutation group P_n. These operations are permutations of labels which can be written as a product of cycles which operate on mutually exclusive sets of labels, e.g.,

$$\begin{pmatrix} 1 & 2 & 3 & 4 & 5 & 6 \\ 2 & 4 & 5 & 1 & 3 & 6 \end{pmatrix} = (1\ 2\ 4)(3\ 5)(6) \tag{2–5}$$

---

[7] R. B. King, *Inorg. Chem.*, **20**, 363, 1981.

The cycle structure of a given permutation in the group P_n can be represented by a sequence of indexed variables, i.e., $x_1x_2x_3$ for the permutation in Equation 2–5. A characteristic feature of the symmetric permutation group P_n for all n is that all permutations having the same cycle structure come from the same conjugacy class.[8] Furthermore, no two permutations with different cycle structures can belong to the same conjugacy class. Therefore, for the symmetric permutation group P_n (but not necessarily for any of its subgroups), the cycle structures of permutations are sufficient to define their conjugacy classes. Furthermore, the number of conjugacy classes of the symmetric group P_n corresponds to the number of different partitions of n, where a *partition of n* is defined as a set of positive integers $i_1, i_2, \ldots, i_k$ whose sum is n (Equation 2–6).

$$\sum_{j=1}^{k} i_j = n \tag{2–6}$$

An alternative presentation of conjugacy class information for the symmetric groups P_n is given by their cycle indices.[9,10,11] A *cycle index* $Z(P_n)$ for a symmetric permutation group P_n is a polynomial of the following form:

$$Z(P_n) = \sum_{i=1}^{i=c} a_i x_1^{c_{i1}} x_2^{c_{i2}} \ldots x_n^{c_{in}} \tag{2–7}$$

In Equation 2–7 c = number of classes (i.e., partitions of n by Equation 2–6), a_i = number of operations of P_n in class i, x_j = dummy variable referring to cycles of length j, and c_{ij} = exponent indicating the number of cycles of length j in class i. These parameters in the cycle indices of the symmetric groups P_n must satisfy the following relationships:

1. Each of the $n!$ permutations of P_n must be in some class, i.e.,

$$\sum_{i=1}^{i=c} a_i = n! \tag{2–8}$$

2. Each of the n ligands must be in some cycle of each permutation in P_n (counting, of course, "fixed points" of cycles of length 1 represented by $x_1^{c_1}$), i.e.,

---

[8]C. D. H. Chisholm, *Group Theoretical Techniques in Quantum Chemistry*, Academic Press, New York, 1976, chap. 6.

[9]G. Pólya, *Acta Math.*, **68**, 145, 1937.

[10]G. Pólya and R. C. Reed, *Combinatorial Enumeration of Groups, Graphs, and Chemical Compounds*, Springer-Verlag, New York, 1987.

[11]N. G. Debruin in *Applied Combinatorial Mathematics*, E. F. Beckenbach, Ed., John Wiley & Sons, New York, 1964, chap. 5.

$$\sum_{j=1}^{j=n} jc_{ij} = n \text{ for } 1 \leq i \leq c \qquad (2\text{–}9)$$

Let us now consider some of the properties of the specific symmetric permutation groups. Any permutation group P_n has a normal subgroup of index 2 (and thus of order $n!/2$), consisting of only the permutations of even parity (which necessarily includes the identity permutation). This special subgroup of P_n is called the *alternating group* and is designated as A_n. The permutation group P_4 has the normal chain $C_1 \triangleleft C_2 \triangleleft D_2 \triangleleft A_4 \triangleleft P_4$, with the respective quotient groups $C_2/C_1 = C_2$, $D_2/C_2 = C_2$, $A_4/D_2 = C_3$, and $P_4/A_4 = C_2$. The permutation group P_4 can thus be expressed as the direct product $C_2 \times C_2 \times C_3 \times C_2$. The permutation group P_4 is isomorphic to the full tetrahedral group T_d, whereas its normal subgroup A_d of index 2 corresponds to the pure rotational subgroup T. The only normal subgroup of the permutation groups P_n $(n \geq 5)$ is the corresponding alternating group A_n. However, A_n $(n \geq 5)$ has no normal subgroups. Therefore, A_n $(n \geq 5)$ is a simple group and cannot be expressed as a direct product of cyclic subgroups.

Permutation groups can also be defined in terms of their generators and relations between their generators.[12] Thus, certain operations in a group, s_1, $s_2, \ldots, s_m$ are called *generators* if every operation in the group is expressible as a finite product of their powers (including negative powers). The cyclic groups C_n have a single generator (i.e., $m = 1$). A set of relations satisfied by the generators of a group is called an *abstract definition* or *presentation* of the group, if every relation satisfied by the generators is an algebraic consequence of these particular relations. Thus, if n is finite, $s^n = E$, where E is the identity operation, is an abstract definition of the cyclic group C_n.

The transitive groups of low degrees are particularly significant in the theory of permutation groups. The transitive groups of degrees up to 11 have been tabulated and their properties are given in detail.[13] All of the transitive permutation groups of degree < 7 are listed in Table 2–3. The following points about Table 2–3 are of interest:

1. In the groups C_n, the order is the same as the number of classes; therefore, these groups are Abelian.
2. The groups C_3, C_4, $A_4 \equiv T$, C_5, $A_5 \equiv I$, and A_6 contain only even operations.
3. The group M_5 is a *metacyclic* group of degree 5. For prime n, metacyclic groups of order $n(n + 1)$ are the largest permutation groups of degree n which is a soluble group. A metacyclic group is defined in terms of two generators s and t and the relationships $s^p = t^{p-1} = E$ and

---

[12] H. S. M. Coxeter and W. O. J. Moser, *Generators and Relations for Discrete Groups*, Springer-Verlag, Berlin, 1972.

[13] G. Butler and J. McKay, *Communications in Algebra*, **11**, 863, 1983.

$t^{-1}st = s^r$ where p is a prime (5 in the case of M_5) and r is a primitive root (mod p), which is 2 when $p = 5$.

4. The group P_5 (=S_5) is *not* isomorphic with the full icosahedral group I_h,which has 10 conjugacy classes.[14]
5. The dihedral group D_3 can be a permutation group of either degree 3 or 6. Similarly, the tetrahedral rotation group T of order 24 can be a transitive permutation group of either degree 4 (e.g., the faces of a tetrahedron), where it is the alternating group A_4, or degree 6 (e.g., the edges of a tetrahedron). In addition, the icosahedral rotation group I of order 60 can be a transitive permutation group of either degree 5, where it is the alternating group A_5, or degree 6.

TABLE 2–3
Transitive Permutation Groups of Degrees Less than Seven

Group	Degree	Order	Number of Classes	Special Properties
C_3	3	3	3	Even, Abelian
$D_3 \equiv P_3 (S_3)$	3	6	3	
C_4	4	4	4	Abelian
D_2	4	4	4	Even, Abelian
D_{2d}	4	8	5	
$A_4 \equiv T$	4	12	4	Even
$P_4 (S_4) \equiv T_d$	4	24	5	
C_5	5	5	5	Even, Abelian
D_5	5	10	4	
M_5	5	20	5	Metacyclic
$A_5 \equiv I$	5	60	5	Even, Simple
$P_5 (S_5)$	5	120	7	**Not I_h!**
C_6	6	6	6	Abelian
D_3	6	6	3	
D_6	6	12	3	
$A_4 \equiv T$	6	12	4	Even
	6	18	9	
	6	24	8	
	6	24	5	Even
	6	24	5	
	6	36	9	
	6	36	6	
	6	48	10	
$L(2,5) \equiv I$	6	60	5	
	6	72	9	
	6	120	7	Isomorphic to P_5
A_6	6	360	7	Even, Simple
$P_6 (S_6)$	6	720	11	

[14]R. B. King and D. H. Rouvray, *Theor. Chim. Acta*, **69**, 1, 1986.

2.6 FRAMEWORK GROUPS

Symmetry point groups specify the symmetry properties of a general three-dimensional object of finite extension and apply to all such objects. However, if the object consists of only a finite number of points, such as the vertices of a polyhedron, then further classification of symmetry by means of groups becomes possible. A three-dimensional object consisting of a finite number of distinct points may be called a *framework*, and its set of symmetry operations then constitutes a *framework group*.[15] Framework groups are of considerable chemical interest, since the points in the framework can, of course, represent atoms in a molecule.

In order to describe a framework group, the relevant symmetry point group is first found and described by its Schoenflies symbol. The distinct points in the framework are then described in terms of their positions relative to subspaces corresponding to the symmetry elements in the group. These subspaces can be classified by their dimensionalities as follows:

1. *0-dimensional:* A central point defining an improper rotation S_n, namely the intersection of a proper rotation axis (C_n) with a perpendicular reflection plane (σ_h). In the case of $n = 2$, the central point is an inversion center (i). Central points are given the generic designation O.
2. *1-dimensional:* A rotation axis (C_n) designated by C_n where n is the order of the rotation.
3. 2-dimensional: A reflection plane (σ) designated by σ_h, σ_v σ_d, depending upon its location in the point group.
4. *3-dimensional:* The remaining part of three-dimensional space external to any of the symmetry elements of the point group, designated as X.

The location of any given site in the framework is specified in terms of the subspace of the *lowest* possible dimensionality. This leads to the preference order $O > C_n > \sigma > X$. Note the correspondence of these subspaces to the symmetry elements in Table 2–1.

Again, the trigonal bipyramid can be used as an example. The framework group of the trigonal bipyramid can be expressed as $D_{3h}[C_3(L_2), 3C_2(L)]$, in which $C_3(L_2)$ means that the two axial vertices are located on the C_3 axis, and $3C_2(L)$ means that each of the equatorial sites is located on a different C_2 axis. The D_{3h} group for the 12 permutations of the five vertices of the trigonal bipyramid is an intransitive permutation group, since the two axial sites are never interchanged with the three equatorial sites.

The property of chirality can be used to classify framework groups. Thus, an object is *chiral* if its symmetry group contains no improper rotations S_n

---

[15] J. A. Pople, *J. Am. Chem. Soc.*, **102**, 4615, 1980.

(including $S_1 \equiv \sigma$ and $S_2 \equiv i$) . Systems with symmetry groups containing one or more improper rotations are thus *achiral.* Framework groups can be classified in terms of their chirality as follows:[16]

1. **Linear.** Framework groups in which all sites are located in a straight line, i.e., in a one-dimensional subspace of three-dimensional space.
2. **Planar.** Non-linear framework groups in which all sites are located in a (flat) plane, i.e., in a two-dimensional subspace of three-dimensional space.
3. **Achiral.** Non-planar framework groups in which the point group contains at least one improper rotation S_n ($n \geq 1$) including $S_1 \equiv \sigma$ or $S_2 \equiv i$.
4. **Chiral.** Non-planar framework groups in which the point group contains no improper rotations.

Consider the use of a framework group to describe the symmetry of a molecule of the type ML_n in which M is a metal or other central atom and the n ligands L may or may not be equivalent but cannot themselves be chiral. The framework group is used to describe the symmetry of the skeleton consisting of the n sites for the ligands L. In general, the ligands L in such an ML_n molecule are not equivalent, so that a *ligand partition*, as well as the underlying *site partition* of the underlying skeleton, may be represented by symbols of the type $(a_1^{b_1}, a_2^{b_2}, \ldots, a_k^{b_k})$, in which a_k and b_k are small positive integers and $a_m \geq a_{m+1}$ $(1 \leq m \leq k)$. In this symbol for the ligand partition, there are b_k sets of a_k identical ligands. Thus, a ligand partition (n) refers to an ML_n complex in which all n ligands L are equivalent. In an analogous symbol for a site partition, there are b_k different sets of a_k equivalent sites on the skeleton. Thus, a site partition (n) refers to an ML_n complex in which all n sites are equivalent, corresponding to a transitive permutation or framework group.

Consider now the relationship between the chirality of ML_n molecules and their underlying framework groups. All ML_n molecules having chiral framework groups are chiral regardless of their ligand partitions. One aspect of the study of *chirality algebra*[17,18] is the question of how unsymmetrical a ligand partition of an ML_n derivative with an achiral framework group must be before the resulting molecule is chiral. Such a study of chirality algebra involves relatively advanced group representation theory and is beyond the scope of this book. Even though planar framework groups have a plane of symmetry containing all of the sites, ML_n derivatives having planar framework groups can sometimes be chiral if the plane of symmetry

[16] R. B. King, *Theor. Chim. Acta,* **63**, 103, 1983.

[17] R. B. King, *J. Math. Chem.*, **2**, 89, 1988.

[18] R. B. King in *New Developments in Molecular Chirality*, P. G. Mezey, Ed., Kluwer Academic Publishers, 1991, 13[illegible].

containing all n sites is eliminated by a process called *polarization*, in which the halfspace above the unique molecular plane is conceptually regarded as positive and the corresponding halfspace below the unique molecular plane is conceptually regarded as negative. The simplest example of such a polarization process involves the conversion of a planar polygon framework group $D_{nh}[nC_2(L)]$ into a framework group of the type $C_{nv}[\frac{1}{2}n\sigma_v(L_2)]$ for even n or $C_{nv}[n\sigma_v(L_2)]$ for odd n. This is the mathematical analogue to the chemical process of symmetrically bonding a planar aromatic hydrocarbon C_nH_n ($3 \leq n \leq 8$) to a transition metal.

Only relatively few framework groups are transitive. The transitive framework groups having 3 to 7 sites are listed in Table 2–4 in order of decreasing symmetry for a given number of sites. The following points concerning the framework groups listed in Table 2–4 are of interest:

1. All of the planar framework groups listed in Table 2–4 can be polarized to give a framework group with half the number of operations.
2. The framework groups for the $D_{3h}[3C_2(L_2)]$ trigonal prism and the $D_{3d}[3\sigma_d(L_2)]$ trigonal antiprism are permutationally equivalent.

TABLE 2–4
Transitive Framework Groups with 7 or Less Sites

Name	Symbol	Sites	Order	Type
Equilateral triangle	$D_{3h}[3C_2(L)]$	3	6	Planar
Regular tetrahedron	$T_d[4C_3(L)]$	4	24	Achiral
Planar square	$D_{4h}[2C_2(L_2)]$	4	16	Planar
Planar rectangle	$D_{2h}[\sigma(L_4)]$	4	8	Planar
Allene (bisphenoid) skeleton	$D_{2d}[2\sigma_d(L_2)]$	4	8	Achiral
Planar pentagon	$D_{5h}[5C_2(L)]$	5	20	Planar
Regular octahedron	$O_h[3C_4(L_2)]$	6	48	Achiral
Planar hexagon	$D_{6h}[3C_2(L_2)]$	6	24	Planar
Trigonal prism	$D_{3h}[3C_2(L_2)]$	6	12	Achiral
Trigonal antiprism	$D_{3d}[3\sigma_d(L_2)]$	6	12	Achiral
Regular heptagon	$D_{7h}[7C_2(L)]$	7	28	Planar

Chapter 3

ATOMIC ORBITALS AND COORDINATION POLYHEDRA

3.1 ATOMIC ORBITALS

Application of ideas from topology and graph theory to chemical bonding first requires consideration of the shape and symmetry of the relevant atomic orbitals. The atomic orbitals correspond to the one-particle wave functions Ψ, obtained as *spherical harmonics* by solution of the following second order differential equation, when the potential energy V is spherically symmetric:

$$\frac{\partial^2\Psi}{\partial x^2} + \frac{\partial^2\Psi}{\partial y^2} + \frac{\partial^2\Psi}{\partial z^2} + \frac{8\pi^2 m}{h^2}(E - V)\Psi = \nabla^2\Psi + \frac{8\pi^2 m}{h^2}(E - V)\Psi = 0 \qquad (3\text{–}1)$$

This equation was known in the 19th century[1] and was used for problems in astronomy involving the motion of celestial bodies. In 1926 Schrödinger[2] recognized the application of the spherical harmonics Ψ obtained by solving Equation 3–1 for the description of the behavior of sub-atomic particles such as electrons. This equation is therefore called the Schrödinger equation.[3] In Equation 3–1 the spherical harmonics Ψ are the so-called wave functions, x, y, and z are the coordinates in space, m is the mass, h is Planck's constant, E is the total energy, and V is the potential energy.

The spherical harmonics Ψ obtained by solving Equation 3–1 are functions of either the three spatial coordinates x, y, and z or the corresponding *spherical polar coordinates* r, θ, and ϕ defined by the equations

$$x = r \sin\theta \cos\phi \qquad (3\text{–}2a)$$

$$y = r \sin\theta \cos\phi \qquad (3\text{–}2b)$$

$$z = r \cos\theta \qquad (3\text{–}2c)$$

For real waves, the wave function Ψ corresponds to the amplitude of the wave and thus has no physical reality when applied to chemical problems.

---

[1]W. E. Byerly, *An Elementary Treatise on Fourier's Series and Spherical, Cylindrical, and Ellipsoidal Harmonics*, Dover, New York, 1959 (reprinted from an 1893 edition published by Ginn and Company).

[2]D. B. Beard and G. B. Beard, *Quantum Mechanics with Applications*, Allyn and Bacon, Boston, 1970, chap. 4.

[3]J. E. Huheey, *Inorganic Chemistry: Principles of Structure and Reactivity*, 3rd Ed. Harper & Row, New York, 1983.

However, the quantity $|\Psi^2|$ corresponds to the probability of finding a particle, such as an electron.

The wave functions Ψ are most conveniently expressed as functions of the spherical polar coordinates (Equations 3–2), since a set of linearly independent wave functions can be found such that Ψ can be factored into the following product:

$$\Psi(r,\theta,\phi) = R(r)\cdot\Theta(\theta)\cdot\Phi(\phi) \tag{3–3}$$

in which the factors R, Θ, and Φ are functions solely of r, θ, and ϕ, respectively. The product $\Theta(\theta)\cdot\Phi(\phi) = A(\theta, \phi)$ is called the *angular component* of the wave function Ψ, since it depends only upon the angular coordinates θ and ϕ whereas $R(r)$ is called the *radial component* of the wave function Ψ, since it depends only upon the radial coordinate r. Since the value of the radial component $R(r)$ of Ψ is completely independent of the angular coordinates θ and ϕ, it is independent of direction (i.e., *isotropic*) and therefore remains unaltered by any symmetry operations. For this reason, all of the symmetry properties of a spherical harmonic Ψ, and thus of the corresponding wave function or atomic orbital, are contained in its angular component A. We can therefore classify spherical harmonics and their corresponding atomic orbitals by the symmetry of their angular components.

Spherical harmonics and therefore the corresponding wave functions or atomic orbitals can also be classified by a series of small integers, which for wave functions or atomic orbitals are called *quantum numbers*. Each of the three factors of Ψ generates a quantum number. Thus the factors $R(r)$, $\Theta(\theta)$, and $\Phi(\phi)$ generate the quantum numbers n, l, and m_l (or simply m), respectively. The *principal quantum number n*, derived from the radial component $R(r)$, relates to the distance from the center of the sphere (i.e., the nucleus in the case of atomic orbitals). The *azimuthal quantum number l*, derived from the factor $\Theta(\theta)$ in Equation 3–3, relates to the number of nodes in the angular component A, where a *node* is a plane corresponding to a zero value of A or Ψ, i.e., where the sign of A changes from positive to negative. Atomic orbitals where $l = 0, 1, 2,$ and 3 have 0, 1, 2, and 3 nodes, respectively, and are conventionally designated as s, p, d, and f orbitals, respectively. For a given value of the azimuthal quantum number l, the *magnetic quantum number* m_l or m, derived from the factor $\Phi(\phi)$ in Equation 3–3, may take on all $2l + 1$ different values from $+l$ to $-l$. There are, therefore, necessarily $2l + 1$ distinct orthogonal orbitals for a given value of l corresponding to 1, 3, 5, and 7 distinct s, p, d, and f orbitals, respectively. The angular components for the most general forms of s, p, d, and f orbitals have 0, 2, 4, and 6 lobes, respectively, of which half of the lobes correspond to positive values of Ψ and the other half of the lobes correspond to negative values of Ψ. When depicting the angular dependence of atomic orbitals, positive lobes are unshaded and negative lobes are conventionally shaded. Table 3–1 illustrates some of the important properties of s, p, and d orbitals.

The angular component of a spherical harmonic $A(\theta,\phi)$ is also called a *surface harmonic* and is written as $Y_{lm}(\theta,\phi)$ in order to indicate the dependence on the azimuthal and magnetic quantum numbers. These surface harmonics are used in the tensor surface harmonic theory of polyhedral boranes and metal clusters, to be discussed later in this book.

TABLE 3–1
Properties of Atomic Orbitals

Type	Nodes	Polynomial	Angular Function	Appearance
s	0		independent of θ, ϕ	spherically symmetrical
p	1	x	$\sin\theta\cos\phi$	
p	1	y	$\sin\theta\sin\phi$	
p	1	z	$\cos\theta$	
d	2	xy	$\sin^2\theta\sin 2\phi$	
d	2	xz	$\sin\theta\cos\theta\cos\phi$	
d	2	yz	$\sin\theta\cos\theta\sin\phi$	
d	2	x^2-y^2	$\sin^2\theta\cos 2\phi$	
d	2	$2z^2-x^2-y^2$ (abbreviated as z^2)	$(3\cos^2\theta-1)$	

The following points concerning the atomic orbitals depicted in Table 3–1 are of interest:

1. All atomic orbitals are *orthogonal*; i. e., the overlap between different atomic orbitals is zero.
2. The extent of the s orbital (with $l = 0$) is independent of direction, i.e., the s orbital is spherically symmetrical and its wave function Ψ_s is independent of θ and ϕ. In view of this and in view of the uniqueness of the ns orbital for a given value of n, there is no polynomial designation for the s orbital.
3. The p_x, p_y, and p_z orbitals all have the same shape, with their nodes in the yz, xz, and xy planes, respectively. Note how the p orbitals are orthogonal to each other and to the s orbital.
4. There are 5 ($= 2l + 1$ for $l = 2$) distinct and mutually orthogonal d orbitals. In the conventionally used set of five d orbitals, four have the same general shape with four lobes and two nodes. The fifth d orbital, namely the $d_{2z^2-x^2-y^2} = d_{z^2}$ orbital has a unique shape. However, all possible pairs of the d orbitals are orthogonal.

3.2 HYBRIDIZATION OF ATOMIC ORBITALS TO FORM COORDINATION POLYHEDRA

Consider a metal complex of the general type ML_n, in which M is the central metal atom, L_n refers to n ligands surrounding M, and each ligand L is attached to M through a single atom of L. (Such ligands are *unidentate* ligands—polydentate or chelating ligands bonded to a metal through two or more atoms are not considered here). The central metal atom M thus forms a total of n chemical bonds with the n ligands L; i.e., n is the *coordination number* of M. The positions of the n ligands L define the n vertices of a polyhedron, known as the *coordination polyhedron* of ML_n. The combined strengths of the n chemical bonds formed by M to the n ligands L are maximized if the metal valence atomic orbitals overlap to the maximum extent with the atomic orbitals of the ligands L. The available metal valence orbitals may be combined or *hybridized* in such a way to maximize this overlap.

Consider a "light" element of the first row of eight of the periodic table Li→F, such as, boron or carbon. The valence orbitals consist of a single s orbital and the three p orbitals, namely p_x, p_y, and p_z (Table 3–1). This set of valence orbitals is conveniently called an *sp^3 manifold.* In the case of methane, CH_4 (Figure 3–1), the four hydrogen atoms are located at the vertices of a regular tetrahedron surrounding the central carbon atom. The strengths of the four C–H bonds directed towards the vertices of a regular tetrahedron can be maximized if the following linear combinations of the wave functions of the atomic orbitals in the sp^3 manifold are used:

$$\psi_1 = \frac{1}{2}\phi_s + \frac{1}{2}\phi_x + \frac{1}{2}\phi_y + \frac{1}{2}\phi_z \qquad (3\text{–}4a)$$

$$\psi_2 = \frac{1}{2}\phi_s - \frac{1}{2}\phi_x - \frac{1}{2}\phi_y + \frac{1}{2}\phi_z \qquad (3\text{–}4b)$$

$$\psi_3 = \frac{1}{2}\phi_s + \frac{1}{2}\phi_x - \frac{1}{2}\phi_y - \frac{1}{2}\phi_z \qquad (3\text{–}4c)$$

$$\psi_4 = \frac{1}{2}\phi_s - \frac{1}{2}\phi_x + \frac{1}{2}\phi_y - \frac{1}{2}\phi_z \qquad (3\text{–}4d)$$

In Equations 4, the p_x, p_y, and p_z orbitals are abbreviated as x, y, and z, respectively (as will be done elsewhere in this book for clarity). In addition, the hybrid wave functions are represented by ψ, and the component atomic orbitals are represented by ϕ. The four sp^3 hybrid orbitals for methane are shown in Figure 3–1.

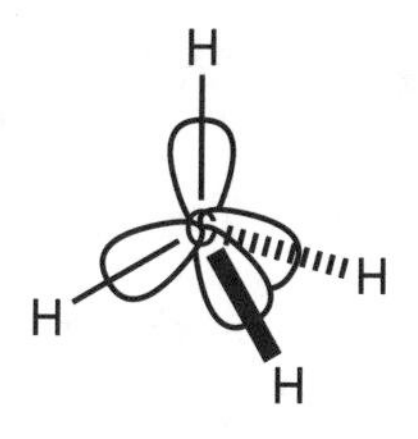

FIGURE 3–1. Methane showing the four sp^3 hybrid orbitals.

The process of determining the coefficients in equations such as 3–4 is beyond the scope of this book and can become complicated when the degrees of freedom are increased by lowering the symmetry of the coordination polyhedron or by increasing the size of the valence orbital manifold to include *d* orbitals. However, elementary symmetry considerations, as outlined in group-theory texts,[4] can be used to determine which atomic orbitals have the necessary symmetry properties to form a hybrid corresponding to a given coordination polyhedron. For example, the four atomic orbitals of an sp^3 manifold can form four hybrid orbitals pointing towards the vertices of a tetrahedron as outlined above. However, the four atomic orbitals of an sp^3 manifold are excluded by symmetry considerations from forming four hybrid orbitals pointing towards the vertices of a planar square or rectangle. Thus, if the plane of the square or rectangle is the xy plane, the p_z orbital (Table 3–1) is seen to have no electron density in this plane (i.e., the xy plane is a node for the p_z orbital) and thus cannot participate in the bonding to atoms in the plane. In the case of coordination polyhedra with larger numbers of vertices, particularly those of relatively high symmetry such as the cube and hexagonal bipyramid for eight-coordination, the inability of certain combinations of atomic orbitals to form the required hybrid orbitals is not as obvious, and more sophisticated group-theoretical methods are required.

These group-theoretical methods rely on the transformation properties of the atomic orbitals in the symmetry point groups of the coordination polyhedra. The transformation properties of an atomic orbital can be described by the irreducible representation of the symmetry point group having characters describing the effects of each symmetry operation on the atomic orbital. Thus a +1 character for a particular symmetry operation means that application of the symmetry operation leaves both the absolute magnitude and sign of the atomic orbital unchanged at every point in three-dimensional space. Similarly, a –1 character for a particular symmetry operation means that its application leaves the absolute magnitude unchanged but changes the sign of the atomic orbital at every point in three-dimensional space. Table 3–2 lists the irreducible representations corresponding to all of the *s*, *p*, and *d*

[4]F. A. Cotton, *Chemical Applications of Group Theory*, John Wiley & Sons, New York, 1971.

TABLE 3–2
Irreducible Representations for the *s*, *p*, and *d* Orbitals

Point Group	Number of Operations	*s* Orbital	*p* Orbitals	*d* Orbitals
C_s	2	A′	A′(*x*,*y*), A″(*z*)	A′(xy,xz,yz,x^2-y^2), A″(z^2)
C_2	2	A	A(*z*), B(*x*,*y*)	A(x^2-y^2,z^2), B(xy,xz,yz)
C_{2v}	4	A_1	$A_1(z)$,$B_1(x)$,$B_2(y)$	$A_1(x^2-y^2,z^2)$, $A_2(xy)$, $B_1(xz)$, $B_2(yz)$
C_{3v}	6	A_1	$A_1(z)$,E(x,y)	$A_1(z^2)$,E(xy,xz,yz,x^2-y^2)
C_{4v}	8	A_1	$A_1(z)$,E(x,y)	$A_1(z^2)$,$B_1(x^2-y^2)$,$B_2(xy)$, E(xz,yz)
D_{2h}	8	A_g	$B_{1u}(z)$,$B_{2u}(y)$, $B_{3u}(x)$	$A_g(x^2-y^2,z^2)$,$B_{1g}(xy)$, $B_{2g}(xz)$,$B_{3g}(yz)$
D_{2d}	8	A_1	$B_2(z)$,E(x,y)	$A_1(z^2)$,$B_1(x^2-y^2)$, $B_2(xy)$,E(xz,yz)
D_{3h}	12	$A_1{'}$	E′(xy),$A_2{''}(z)$	$A_1{'}(z^2)$,E′(x^2-y^2,xy), E″(xz,yz)
D_{3d}	12	A_{1g}	$A_{2u}(z)$,$E_u(x,y)$	$A_{1g}(z^2)$,$E_g(xy,xz,yz,x^2-y^2)$
D_{4h}	16	A_{1g}	$A_{2u}(z)$,$E_u(x,y)$	$A_{1g}(z^2)$,$B_{1g}(x^2-y^2)$, $B_{2g}(xy)$,$E_g(xz,yz)$
D_{4d}	16	A_1	$B_2(z)$,$E_1(x,y)$	$A_1(z^2)$,$E_2(x^2-y^2,xy)$, $E_3(xz,yz)$
D_{5h}	20	$A_1{'}$	$E_1{'}(x,y)$,$A_2{''}(z)$	$A_1{'}(z^2)$,$E_2{'}(x^2-y^2,xy)$, $E_1{''}(xz,yz)$
T_d	24	A_g	$T_u(x,y,z)$	$E_g(z^2,x^2-y^2)$,$T_g(xz,yz,xy)$
O_h	48	A_{1g}	$T_{1u}(x,y,z)$	$E_g(z^2,x^2-y^2)$,$T_g(xz,yz,xy)$

orbitals for the symmetry point groups of the coordination polyhedra of interest. These can be readily found in standard character tables.

The following points can be seen concerning the irreducible representations for the *s*, *p*, and *d* orbitals, as summarized in Table 3–2:

1. The *s* orbitals always belong to the fully symmetrical irreducible representation $A_{(1)(g)}$ having characters of +1 for all operations in the symmetry point group.
2. For point groups having inversion centers (e.g., D_{2h}, D_{4h}, D_{3d}, O_h) the *p* orbitals always belong to ungerade irreducible representations designated by *u* subscripts, since they always change sign upon inversion. Similarly, for these point groups, the *d* orbitals always belong to gerade irreducible representations designated by *g* subscripts, since they do not change sign upon inversion.

The following group-theoretical procedure can be used to determine the irreducible representations for the hybrid orbitals for a given coordination polyhedron with a symmetry point group G in an ML_n complex in which the M–L bonds are anodal σ-bonds from the metal M to a single atom of the ligand L:

1. For each conjugacy class of the symmetry point group G, determine the number of vertices that remain fixed upon applying any of the symmetry operations in the class. This number is the character for any operation in this class of the representation Γ_σ representing the hybrid orbitals for the polyhedron in question. The subscript σ refers to the M–L σ-bonding implied by this procedure. Furthermore, the number of vertices remaining fixed should be the same for any of the operations in a given conjugacy class—if not, a mistake has been made somewhere, probably in visualizing the problem.
2. In general the representation Γ_σ will be a reducible representation. The next step is to reduce Γ_σ into a sum of irreducible representations using character tables and standard group-theoretical procedures based on the great orthogonality theorem.[5]
3. Determine which atomic orbitals correspond to the irreducible representations adding up to Γ_σ. These are the atomic orbitals which are symmetry-allowed for the hybrids.

This procedure can be illustrated by the characters in Table 3–3 for the elementary example of methane CH_4 (Figure 1) treated as an ML_n complex having T_d symmetry in which M = C and L = H. The characters of Γ_σ are derived from the fact that all four vertices of the tetrahedron remain fixed under the identity operation E, each three-fold rotation of C_3 leaves exactly one vertex fixed since each three-fold axis passes through exactly one vertex, each reflection plane σ_d leaves exactly the two vertices fixed that lie in the reflection plane, and each of the remaining symmetry operations (C_2 and S_4) leaves no vertex fixed. The reducible representation Γ_σ can be seen to be the sum of the irreducible representations A_1+T_2, corresponding to the *s* orbital and the three *p* orbitals, respectively, implying that the familiar sp^3 tetrahedral hybrids are allowed by symmetry considerations.

The results of the application of this procedure to most of the familiar (and some less familiar) coordination polyhedra are summarized in Table 3–4.

The only possible coordination polyhedron for coordination number four, using only *s* and *p* orbitals, is the tetrahedron formed by an sp^3 hybrid, as

---

[5]D. Gorenstein, *Finite Groups*, Harper & Row, New York, 1968, chap.3.

TABLE 3–3
The Characters for T_d Methane Corresponding to the Tetrahedral Hybrids

	E	$8C_3$	$3C_2$	$6S_4$	$6\sigma_d$
G_s	4	1	0	0	2
$A_1(s)$	1	1	1	1	1
$T_2(x,y,z)$	3	0	–1	–1	1

noted above. The following comments[6] can be made about coordination numbers five through nine, based on a central sp^3d^n-hybridized metal atom and the information presented in Table 3–4:

Coordination number 5: The two possible coordination polyhedra are the trigonal bipyramid where $\Gamma_\sigma = 2A_1' + E' + A_2''$ corresponding to $sp^3d(z^2)$ hybrids and the square pyramid where $\Gamma_\sigma = 2A_1 + B_1 + E$ corresponding to $sp^3d(x^2-y^2)$ hybrids depicted below. Both of these coordination polyhedra are found in five-coordinate ML_5 metal complexes.[7]

Square Pyramid

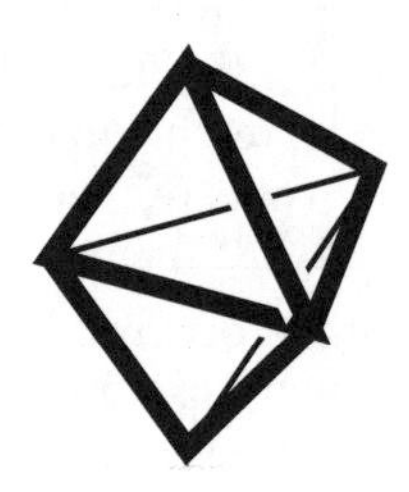

Trigonal Bipyramid

Coordination number 6:[8] The very symmetrical O_h octahedron, in which $\Gamma_\sigma = A_{1g} + E_g + T_{1u}$ corresponding to $sp^3d^2(x^2-y^2,z^2)$ hybridization, is overwhelmingly favored for this coordination number. The D_{3h} trigonal prism, in which $\Gamma_\sigma = A_1' + E' + A_2'' + E''$ corresponding to $sp^3d^2(xz,yz)$ hybridization, is found for tris(ethylenedithiolate) derivatives of early and middle transition metals,[9] and a distorted pentagonal pyramid, in which $\Gamma_\sigma = 2A_1 + E_1 + E_2$ corresponding to $sp^3d^2(xy,x^2-y^2)$ hybridization, is found for certain complexes of post-transition elements such as tellurium[10] and

---

[6]R. B. King, *J. Am. Chem. Soc.*, **91**, 7211, 1969.
[7]E. L. Muetterties and R. A. Schunn, *Quart. Rev. (London)*, **20**, 245, 1966.
[8]D. L. Kepert, *Prog. Inorg. Chem.*, **23**, 1, 1977.
[9]R. Eisenberg, *Prog. Inorg. Chem.*, **12**, 295, 1970.
[10]R. B. King, S. A. Sangokoya, and E. M. Holt, *Inorg. Chem.*, **26**, 4307, 1987.

antimony.[11] Note that the choice of the two *d* orbitals to be used in the sp^3d^2 hybrid determines which of the six-coordinate polyhedra is formed.

TABLE 3–4
The Irreducible Representations for the Hybrid Orbitals Corresponding to Coordination Polyhedra

Polyhedron	G	ϖ	e	f	Γ_σ
Tetrahedron	T_d	4	6	4	$A_1(s)+T_2(x,y,z)$
Square pyramid	C_{4v}	5	8	5	$2A_1(s,z,z^2)+B_1(x^2-y^2)$ $+E(x,y,xz,yz)$
Trigonal bipyramid	D_{3h}	5	9	6	$2A_1'(s,z^2)+E'(x,y,x^2-y^2,xy)$ $+A_2''(z)$
Trigonal prism	D_{3h}	6	9	5	$A_1'(s,z^2)+E'(x,y,x^2-y^2,xy)$ $+A_2''(z)+E''(xz,yz)$
Pentagonal pyramid	C_{5v}	6	10	6	$2A_1(s,z,z^2)+E_1(x,y,xz,yz)$ $+E_2(x^2-y^2,xy)$
Octahedron	O_h	6	12	8	$A_{1g}(s)+E_g(z^2,x^2-y^2)+T_{1u}(x,y,z)$
Capped octahedron	C_{3v}	7	15	10	$3A_1(s,z,z^2)$ $+2E(x,y,x^2-y^2,xy,xz,yz)$
Pentagonal bipyramid	D_{5h}	7	15	10	$2A_1'(s,z^2)+E_1'(x,y)$ $+E_2'(x^2-y^2,xy)+A_2''(z)$
4-Capped trigonal prism	C_{2v}	7	13	8	$3A_1(s,z,z^2,x^2-y^2)$ $+A_2(xy)+2B_1(x,xz)+B_2(y,yz)$
Cube	O_h	8	12	6	$A_{1g}(s)+T_{2g}(xy,xz,yz)+$ A_{2u}(no s,p,d!)$+T_{1u}(x,y,z)$
3,3-Bicapped trigonal prism	D_{3h}	8	15	9	$2A_1'(s,z^2)+E'(x,y,x^2-y^2,xy)$ $+2A_2''(z)+E''(xz,yz)$
Square antiprism	D_{4d}	8	16	10	$A_1(s,z^2)+B_2(z)+E_1(x,y)$ $+E_2(x^2-y^2,xy)+E_3(xz,yz)$
Bisdisphenoid ("D_{2d} dodecahedron")	D_{2d}	8	18	12	$2A_1(s,z^2)+2B_2(z,xy)$ $+2E(x,y,xz,yz)$
Tricapped trigonal prism	D_{3h}	9	21	14	$2A_1'(s,z^2)+2E'(x,y,x^2-y^2,xy)$ $+A_2''(z)+E''(xz,yz)$
Capped square antiprism	C_{4v}	9	20	13	$3A_1(s,z,z^2)+B_1(x^2-$ $y^2)+B_2(xy)+2E(x,y,xz,yz)$

[11]M. J. Begley, D. B. Sowerby, and I. Haiduc, *J. Chem. Soc. Dalton,* 145, 1987.

Octahedron **Trigonal Prism** **Pentagonal Pyramid**

Coordination number 7:[12] The seven-coordinate polyhedron of maximum symmetry is the D_{5h} pentagonal bipyramid, in which $\Gamma_\sigma = 2A_1' + E_1' + E_2' + A_2''$ corresponding to $sp^3d^3(xy,x^2-y^2,z^2)$ hybridization. This polyhedron is commonly found in seven-coordinate complexes. Other polyhedra found in seven-coordinate complexes include the capped octahedron with $\Gamma_\sigma = 3A_1 + 2E$ corresponding to $sp^3d^3(z^2,xz,yz)$ or $sp^3d^3(z^2,xy,x^2-y^2)$ hybridization and the 4-capped trigonal prism with $\Gamma_\sigma = 3A_1 + A_2 + 2B_1 + B_2$ corresponding to $sp^3d^3(z^2,xy,xz)$ or $sp^3d^3(x^2-y^2,xy,xz)$ hybridization. Note that the choice of *d* orbitals again affects the choice of coordination polyhedron.

Pentagonal Bipyramid **Capped Octahedron**

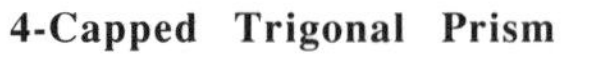

4-Capped Trigonal Prism

Coordination Number 8:[13,14,15] The common coordination polyhedra for coordination number eight are the bisdisphenoid or "D_{2d} dodecahedron" with $\Gamma_\sigma = 2A_1 + 2B_2 + 2E$ corresponding to $sp^3d^4(z^2,xy,xz,yz)$ hybridization and the square antiprism with $\Gamma_\sigma = A_1 + B_2 + E_1 + E_2 + E_3$ corresponding to $sp^3d^4(x^2-y^2,xy,xz,yz)$ hybridization.

---

[12] D. L. Kepert, *Prog. Inorg. Chem.*, **25**, 41, 1979.

[13] E. L. Muetterties and C. M. Wright, *Quart. Rev. (London)*, **21**, 109, 1967.

[14] S. J. Lippard, *Adv. Inorg. Chem.*, **8**, 109, 1967.

[15] D. L. Kepert, *Prog. Inorg. Chem.*, **24**, 179, 1978.

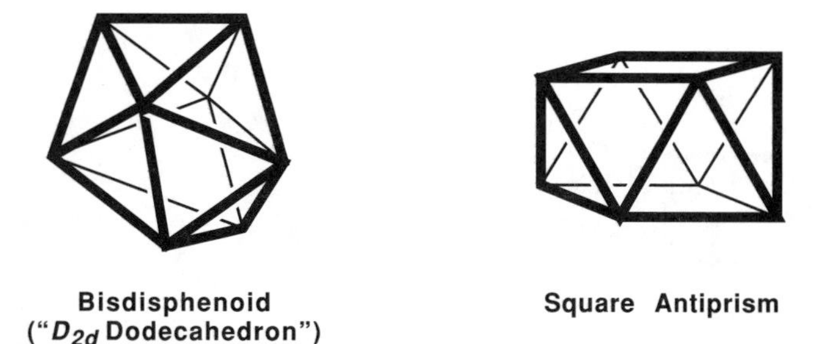

Bisdisphenoid
("D_{2d} Dodecahedron")

Square Antiprism

Three eight-vertex polyhedra of relatively high symmetry, namely the cube, the hexagonal bipyramid, and the D_{3h} 3,3-bicapped trigonal prism, cannot be formed using solely *s*, *p*, and *d* atomic orbitals. Such polyhedra with nine or fewer vertices, which cannot be formed from a sp^3d^5 nine-orbital manifold, are called *forbidden coordination polyhedra*, or more specifically, *spd-forbidden coordination polyhedra.*[16] All of these eight-vertex polyhedra have symmetry point groups that can be expressed as direct products $R \times C_s'$. For the cube and hexagonal bipyramid, the group $C_s' = C_i$ and $R = O$ and D_6, respectively, so that the inversion center *i* is the primary involution (Section 2.4), whereas for the 3,3-bicapped trigonal prism, the group $C_s' = C_s$ and $R = D_3$ so that the horizontal reflection plane σ_h is the primary involution. The character of the primary involution of the reducible representation corresponding to Γ_σ is equal to the number of vertices which remain fixed when the primary involution is applied. If the primary involution is an inversion, its character is necessarily zero since *no* vertices of a polyhedron remain fixed by an inversion. Therefore, the reducible representation Γ_σ of an 8-vertex polyhedron with an inversion center contains equal numbers of even and odd irreducible representations. This corresponds to a hybridization using four symmetrical and four antisymmetrical atomic orbitals. Since only three orbitals of the sp^3d^5 manifold are antisymmetrical (namely the three *p* orbitals), an eight vertex polyhedron with an inversion center cannot be formed using only *s*, *p*, and *d* orbitals.

---

[16]R. B. King, *Theor. Chim. Acta*, **64**, 453, 1984.

Cube **Hexagonal Bipyramid** **3,3-Bicapped Trigonal Prism**

Coordination number 9: The deltahedron of maximum symmetry is the D_{3h} tricapped trigonal prism with $\Gamma_\sigma = 2A_1' + 2E' + A_2'' + E''$. Either this nine-vertex polyhedron or the nine-vertex capped square antiprism with $\Gamma_\sigma = 3A_1 = B_1 + B_2 + 2E$ can be formed using only a nine-orbital sp^3d^5 manifold and thus are feasible nine-vertex coordination polyhedra. The small number of nine-coordinate complexes including the hydrides[17,18] ReH_9^{2-} and TcH_9^{2-} generally use the tricapped trigonal prism.

Tricapped Trigonal Prism

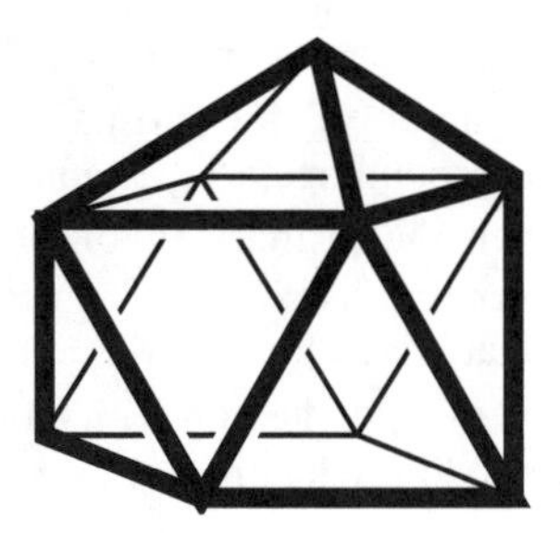

Capped Square Antiprism

[17] K. Knox and A. P. Ginsberg, *Inorg. Chem.*, **3**, 555, 1964.

[18] S. C. Abrahams, A. P. Ginsberg, and K. Knox, *Inorg. Chem.*, **3**, 559, 1964.

Chapter 4

DELOCALIZATION IN HYDROCARBONS AND BORANES

4.1 GRAPH THEORY AND HÜCKEL THEORY

The early theoretical treatment of chemical bonding featured a rivalry between the *valence bond theory* and *molecular orbital theory*.[1] In valence bond theory all bonds are considered to be *localized*, i.e., two-center bonds involving overlap of only two atomic orbitals, which are located on adjacent atoms. Molecular orbital theory facilitates the study of delocalization through interactions of the two-center bonds of valence bond theory or, equivalently, the atomic orbitals forming the bonds leading ultimately to what may be regarded as n-center bonds where $n > 2$; these bonds are described by *molecular orbitals* delocalized over the n atoms of a n-center bond. A *delocalized* chemical bond can thus be defined as a chemical bond involving the interactions of three or more atomic orbitals. Such interactions involve neighborhood relationships of the relevant atomic orbitals and are therefore readily described by the methods of graph theory and topology. Stabilization by delocalization is traditionally called "aromaticity" in view of its original application in understanding the stability of benzene, an "aromatic" molecule because of its odor. This chapter discusses delocalized chemical bonds in structures containing atoms having only s and p orbitals in the valence orbital manifold, namely carbon and boron. The carbon structures of interest are the hydrocarbons of organic chemistry, particularly planar "aromatic" hydrocarbons such as the prototypical benzene. The boron structures of interest are the deltahedral boranes of inorganic chemistry. The ideas presented in this chapter thus illustrate how some key ideas in organic chemistry can be applied to inorganic chemistry problems.

A frequently used approach for the study of delocalized chemical bonding uses the linear combination of atomic orbitals (LCAO) to form molecular orbitals. A linear combination of n atomic orbitals thus leads to n molecular orbitals, some of which are bonding and others antibonding. In this context, a *bonding* molecular orbital is one in which the presence of electrons stabilizes the chemical bond, whereas an *antibonding* molecular orbital is one in which the presence of electrons destabilizes the chemical bond. Each molecular orbital is written as a linear combination of the overlapping atomic orbitals on the atoms participating in the delocalized bonding leading to the n equations ($1 \leq k,i \leq n$)

---

[1]J. E. Huheey, *Inorganic Chemistry: Principles of Structure and Reactivity*, 3rd Ed., Harper & Row, New York, 1983, 92-100.

$$\psi_k = \sum_i c_{ik}\phi_i \tag{4–1}$$

in which ψ_k is the wave function of the kth molecular orbital and ϕ_i is the wave function of the ith atomic orbital. The ϕ_i atomic orbital wave functions are sometimes called a *basis set* and are conventionally normalized by using the equation

$$\int \phi_i \phi_j \, d\tau = 1 \tag{4–2}$$

The wave equation is written in a special form, called the *secular equation* as follows:

$$\mathcal{H}\psi = E\psi \Rightarrow \mathcal{H}\psi - E\psi = (\mathcal{H} - E)\psi = 0 \tag{4–3}$$

Combining Equation 4–1 for ψ with Equation 4–3 gives the following set of n simultaneous equations:

$$\sum_i c_i (\mathcal{H} - E)\phi_i = 0 \tag{4–4}$$

Multiplication of an equation of the type 4–4 by $\phi_j \, (1 \leq j \leq n)$ followed by integration of the left side over all spatial coordinates of the wave function gives a total of n^2 simultaneous equations of the type

$$\sum_i c_{ik} \int \phi_i (\mathcal{H} - E)\phi_i d\tau = 0 \tag{4–5}$$

Now define the integrals by the following parameters:

$$\alpha_{ii} = \int \phi_i \, \mathcal{H} \phi_i \, d\tau \qquad \text{(energy of the atomic orbital } \phi_i\text{)} \tag{4–6a}$$

$$\beta_{ij} = \int \phi_i \, \mathcal{H} \phi_{ij} \, d\tau \qquad \text{(energy of the interactions between pairs of atomic orbitals)} \tag{4–6b}$$

$$S_{ij} = \int \phi_i \phi_j \, d\tau \qquad \text{(overlap integrals)} \tag{4–6c}$$

The Hückel approximation estimates these integrals as follows:

$$\alpha_{ii} = \alpha \text{ for all orbitals of the same type on the same atom} \tag{4–7a}$$

$$\beta_{ij} = \beta \text{ for all equivalent adjacent orbital-orbital interactions} \tag{4–7b}$$

$$\beta_{ij} = 0 \text{ for all pairs of non-adjacent orbitals} \tag{4–7c}$$

$$S_{ij} = 0 \text{ for all } i, j \tag{4–7d}$$

Methods derived from topology and graph theory can be used to describe the neighborhood relationships of the atomic orbitals interacting to form molecular orbitals.[2,3,4,5] First write the n^2 simultaneous equations indicated in Equation 4–5 in determinant form as follows:

$$|\mathbf{H} - E\mathbf{S}| = 0 \tag{4–8}$$

In Equation 4–8, **H** and **S** are $n \times n$ matrices, which are called the *energy* and *overlap* matrices, respectively.

The overlap of the atomic orbitals involved in delocalized bonding can be represented by a graph G in which the v vertices correspond to the orbitals and the e edges correspond to orbital overlaps. This bonding described by the graph G may be called the *skeletal* bonding of the molecule, with the graph G describing the molecular skeleton. As noted in Chapter 1, the adjacency matrix **A** of G is a $v \times v$ matrix defined as

$$A_{ij} = \begin{cases} 0 \text{ if i = j} \\ 1 \text{ if i and j are connected by an edge} \\ 0 \text{ if i and j are not connected by an edge} \end{cases} \tag{4–9}$$

The eigenvalues of **A** are obtained by the determinantal equation

$$|x\mathbf{I} - \mathbf{A}| = 0 \tag{4–10}$$

in which **I** is the unit matrix defined by $I_{ii} = 1$ and $I_{ij} = 0$ for $i \neq j$. The definitions of α, β, and S in Equations 4–7 can be applied to Equation 4–8 to resolve the energy matrix **H** and the overlap matrix **S** into the unit matrix **I** and the adjacency matrix **A** as follows:

$$\mathbf{H} = \alpha\mathbf{I} + \beta\mathbf{A} \tag{4–11a}$$

$$\mathbf{S} = \mathbf{I} + S\mathbf{A} \tag{4–11b}$$

The energy levels of the Hückel molecular orbitals (Equation 4–8) are thus related to the eigenvalues, x_k, of the adjacency matrix **A** by the following equation, where α, β, and S are defined by Equations 4–7:

---

[2] K. Ruedenberg, *J. Chem. Phys.*, **22**, 1878, 1954.

[3] H. H. Schmidtke, *J. Chem. Phys.*, **45**, 3920, 1966.

[4] H. H. Schmidtke, *Coord. Chem. Rev.*, **2**, 3, 1967.

[5] I. Gutman and N. Trinajstić, *Top. Curr. Chem.*, **42**, 49, 1973.

$$E_k = \frac{\alpha + x_k\beta}{1 + x_kS} \tag{4–12}$$

Application of Equation 4–12 for the determination of energy parameters of bonding and antibonding molecular orbitals requires determination of the three Hückel parameters α, β, and S in order to relate the eigenvalues x_k to the corresponding Hückel molecular orbital energy parameters E_k. The qualitative graph-theory derived methods consider only the topological contributions to molecular orbital energies so that Equation 4–12 is reduced to

$$E_k \approx x_k\beta \tag{4–13}$$

in which E_k is measured relative to the center point α, S is taken to be zero (i.e., Equation 4–7d), and β is an energy unit derivable from experimental data or from molecular orbital energies obtained by some other computational method. Reduction of Equation 4–12 to Equation 4–13 by setting S equal to zero implies that the energy parameters E_k are directly proportional to the eigenvalues x_k of the adjacency matrix **A**. As long as S is zero or positive, positive values of x_k correspond to bonding orbitals and negative values of x_k correspond to antibonding orbitals.

4.2 VERTEX ATOMS

The atoms at the vertices of a polygonal or polyhedral molecule can be classified as light atoms or heavy atoms. A *light atom*, such as boron or carbon, uses only its s and p orbitals for chemical bonding and therefore has the four valence orbitals of its sp^3 manifold. A *heavy atom*, such as a transition metal or post-transition element, uses s, p, and d orbitals for chemical bonding and therefore has the nine valence orbitals of its sp^3d^5 manifold. A *normal vertex atom* in a polygonal or polyhedral molecule uses three of its k^2 valence orbitals ($k = 2$ for a light atom and $k = 3$ for a heavy atom) for the skeletal bonding within the polygon or polyhedron; these three orbitals are called *internal orbitals*. This chapter discusses the skeletal chemical bonding in polygonal and polyhedral molecules with light vertex atoms, namely boron and/or carbon, using the normal three internal orbitals.

The use of three internal orbitals for intrapolygonal or interpolyhedral skeletal bonding by a normal vertex atom leaves one orbital in the case of a light vertex atom or six orbitals in the case of a heavy vertex atom for bonding to an atom or group external to the polygon or polyhedron. Such orbitals are called *external orbitals*. The single external orbital of a light vertex atom can bond to a single monovalent external group such as hydrogen, halogen, alkyl, aryl, alkoxy, dialkylamino, nitro, cyano, etc. This relates to the stoichiometries C_nH_n, $B_nH_n^{2-}$, and $C_2B_{n-2}H_n$ for the planar polygonal hydrocarbons, the cage borane dianions, and the cage carboranes, respectively.

In almost all polygonal and polyhedral molecules, each vertex atom has the electronic configuration of the next rare gas, which is neon in the case of the light vertex atoms boron and carbon. As a result of this rule, each external orbital of the vertex atom must be filled by an electron pair, with the electrons coming from the vertex atom and/or an external group. This provides a means for calculating the number of electrons provided by various vertex groups to the polygonal or polyhedral skeleton; such electrons are called *skeletal electrons*. For example, a BH vertex in polyhedral borane functions is a donor of two skeletal electrons, determined as follows:

Boron valence electrons =	3 electrons
Less: boron electron required for external B–H bond	–1 electron
Skeletal electrons from a BH vertex	2 electrons

Similarly, a CH vertex in a polygonal hydrocarbon or polyhedral carborane functions as a donor of three skeletal electrons. The external hydrogen atoms in BH and CH vertices can be replaced by other monovalent groups without affecting the number of skeletal electrons donated by the vertex in question.

The two extreme types of skeletal chemical bonding in molecules formed by polygonal or polyhedral clusters of atoms, including planar aromatic hydrocarbons and polyhedral boranes (as well as various types of metal clusters to be treated in later chapters), can be called *edge-localized* and *globally delocalized*.[6,7,8,9] An edge-localized polygon or polyhedron has two-electron two-center bonds along each edge. A globally delocalized polygon or polyhedron has a multicenter bond involving all of the vertex atoms (hence the adjective "global"); such global delocalization is a feature of fully aromatic systems, whether two-dimensional, such as benzene, C_6H_6, or three-dimensional, such as the deltahedral borane anions $B_nH_n^{2-}$ ($6 \leq n \leq 12$).

Consideration of the properties of vertex groups leads to the following very simple rule to determine whether polygonal or polyhedral molecules exhibit delocalized bonding or edge-localized bonding:

Delocalization occurs when there is a mismatch between the vertex degree of the polygon or polyhedron and the number of internal orbitals provided by the vertex atom.

---

[6]R. B. King and D. H. Rouvray, *J. Am. Chem. Soc.*, **99**, 7834, 1977.

[7]R. B. King in *Chemical Applications of Topology and Graph Theory*, R. B. King, Ed., Elsevier, Amsterdam, 1983, 99.

[8]R. B. King in *Molecular Structure and Energetics*, J. F. Liebman and A. Greenberg, Eds., VCH Publishers, Deerfield Beach, FL, 1976, 123.

[9]R. B. King, *J. Math. Chem.*, **1**, 249, 1987.

This rule is illustrated in Table 4–1 for normal vertex atoms providing three internal orbitals. This rule implies that fully edge-localized bonding occurs in a polyhedral molecule in which all vertices have degree 3. Such is the case for the polyhedranes $C_{2n}H_{2n}$ (Figure 4–1), such as tetrahedrane ($n = 2$), cubane ($n = 4$), and dodecahedrane ($n = 10$); the vertex degrees are all three, which match the three available internal orbitals leading to edge-localized bonding represented by the $\frac{3}{2}n$ two-center carbon-carbon bonds of the skeleton. In the planar polygonal molecules $C_nH_n^{(n-6)+}$ ($n = 5, 6, 7$), the vertex degrees are all two and thus do not match the available three internal orbitals, thereby leading to globally delocalized two-dimensional aromatic systems. Furthermore, polyhedral molecules having all normal vertex atoms are globally delocalized if all vertices of the polyhedron have degrees 4 or larger; the simplest such polyhedron is the regular octahedron with six vertices, all of degree 4. Tetrahedral chambers in deltahedra, which lead to isolated degree 3 vertices, provide sites of localization in an otherwise delocalized molecule, provided, of course, that all vertex atoms use the normal three internal orbitals.

TABLE 4–1
Delocalized Versus Localized Bonding and the "Matching Rule"
(assumes three internal orbitals per vertex atom)

Structure Type	Vertex Degrees	Matching	Localization	Examples
Planar polygons	2	No	Delocalized	Benzene, $C_5H_5^-$, $C_7H_7^+$
"Simple polyhedra"	3	Yes	Localized	Polyhedranes: C_4H_4, C_8H_8, $C_{20}H_{20}$
Deltahedra	4,5 (6)	No	Delocalized	Polyhedral boranes and carboranes

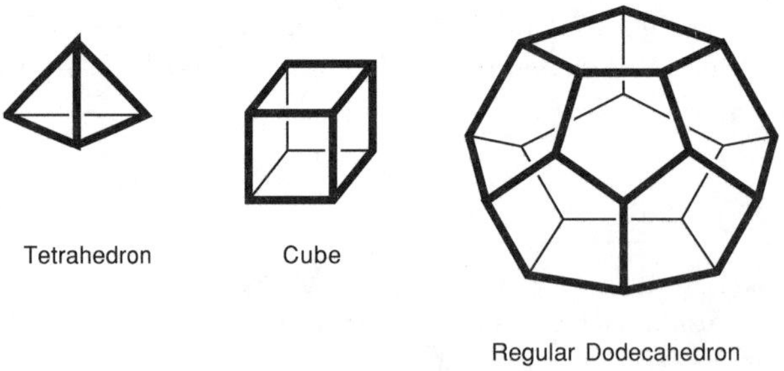

Tetrahedron

Cube

Regular Dodecahedron

FIGURE 4–1. The degree 3 vertex regular polyhedra found in tetrahedrane, C_4H_4; cubane, C_8H_8; and dodecahedrane, $C_{20}H_{20}$.

4.3 APPLICATIONS TO GLOBALLY DELOCALIZED HYDROCARBONS, BORANES, AND CARBORANES

A major achievement of the graph-theory derived approach to the chemical bonding topology of globally delocalized systems is the demonstration of the close analogy between the bonding in two-dimensional planar aromatic systems such as benzene and the bonding in three dimensional deltahedral boranes and carboranes. In such systems with n vertices, the three internal orbitals on each vertex atom are partitioned into two twin internal orbitals (called *tangential* in some other methods[10]) and a unique internal orbital (called *radial* in some other methods). Pairwise overlap between the $2n$ twin internal orbitals is responsible for the formation of the polygonal or deltahedral framework and leads to the splitting of these $2n$ orbitals into n bonding and n antibonding orbitals. The magnitude of this splitting can be designated as $2\beta_s$ where β_s refers to the parameter β in Equations 4–11a, 4–12, and 4–13. This portion of the chemical bonding topology can be described by a disconnected graph G_s having $2n$ vertices corresponding to the $2n$ twin internal orbitals and n isolated K_2 components; a K_2 component has only two vertices joined by a single edge. The dimensionality of this bonding of the twin internal orbitals is one less than the dimensionality of the globally delocalized system. Thus, in the case of the two-dimensional planar polygonal systems, the pairwise overlap of the $2n$ twin internal orbitals leads to the σ-bonding network, which may be regarded as a set of one-dimensional bonds along the perimeter of the polygon involving adjacent pairs of polygonal vertices. The n bonding and n antibonding orbitals thus correspond to the σ-bonding and σ^*-antibonding orbitals, respectively. In the case of the three-dimensional deltahedral systems, the pairwise overlap of the $2n$ twin internal orbitals results in bonding over the two-dimensional surface of the deltahedron, which may be regarded as topologically homeomorphic to the sphere.[11]

The equal numbers of bonding and antibonding orbitals formed by pairwise overlap of the twin internal orbitals are supplemented by additional bonding and antibonding orbitals formed by the global mutual overlap of the n unique internal orbitals. The nature of the relevant unique internal orbitals is depicted in Figure 4–2. In the case of the two-dimensional planar polygonal hydrocarbons, such as benzene, the unique internal orbitals on the vertex carbon atoms are the single p_z orbitals, namely the p orbital perpendicular to the plane of the polygon, designated as the xy plane. These p orbitals are *uninodal* orbitals; i.e., they have a single node, namely the node in the xy

---

[10] K. Wade, *Adv. Inorg. Chem. Radiochem.*, *18*, 1, 1976.

[11] M. J. Mansfield, *Introduction to Topology*, Van Nostrand, Princeton, New Jersey, 1963, 40.

plane. The two-dimensional aromaticity in the planar polygonal hydrocarbons can also be called *uninodal orbital aromaticity*, since the orbitals participating in the global mutual overlap have this single node. Similarly, in the case of the three-dimensional deltahedral boranes of the type $B_nH_n^{2-}$ the unique internal orbitals on the vertex boron atoms are *sp* hybrids with the other *sp* hybrid on each boron atom directed towards the atomic orbital of the external hydrogen atom. The three-dimensional aromaticity in the deltahedral boranes can also be called *anodal orbital aromaticity* since the *sp* hybrids functioning as unique internal orbitals have no nodes.

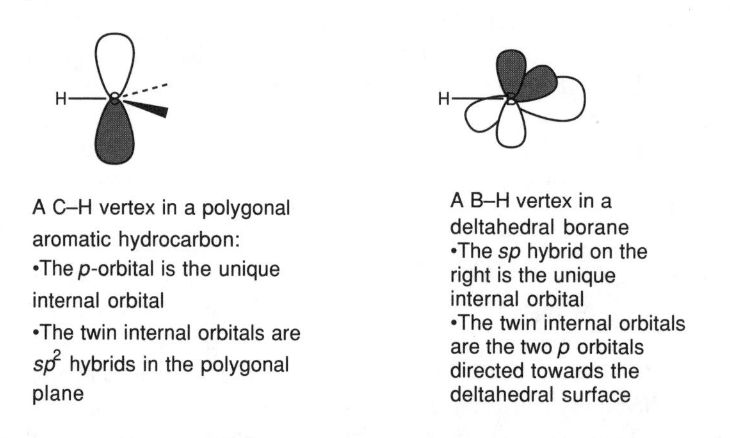

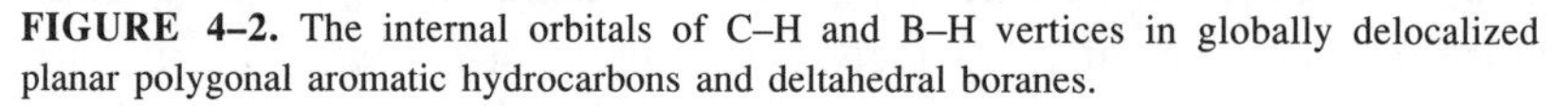

FIGURE 4–2. The internal orbitals of C–H and B–H vertices in globally delocalized planar polygonal aromatic hydrocarbons and deltahedral boranes.

The bonding topology of the *n* unique internal orbitals, whether the uninodal *p* orbitals in the planar polygonal aromatic hydrocarbons or the anodal *sp* hybrids in the three-dimensional deltahedral boranes, can be described by a graph G_c, in which the vertices correspond to the vertex atoms of the polygon or deltahedron, or equivalently their unique internal orbitals, and the edges represent pairs of overlapping unique internal orbitals. The energy parameters of the additional molecular orbitals arising from such overlap of the unique internal orbitals are determined from the eigenvalues of the adjacency matrix $\mathbf{A_c}$ of the graph G_c using β, or more specifically β_c, as the energy unit (Equations 4–11a, 4–12, 4–13). In the case of the two-dimensional aromatic system benzene, the graph G_c is the C_6 cyclic graph (the 1-skeleton[12] of the hexagon), which has three positive (+2, +1, +1) and three negative (–2, –1, –1) eigenvalues, corresponding to the three π-bonding and three π^*-antibonding orbitals, respectively (Figure 4–3). The spectra of the cyclic graphs C_n all have odd numbers of positive eigenvalues,[13] leading

[12]B. Grünbaum, *Convex Polytopes*, Interscience, New York, 1967, 138.
[13]N. L. Biggs, *Algebraic Graph Theory*, Cambridge University Press, London, 1974, 17.

to the familiar $4k + 2$ (k = integer) π-electrons[14] for planar aromatic hydrocarbons. The total benzene skeleton thus has 9 bonding orbitals (6σ and 3π) which are filled by the 18 skeletal electrons which arise when each of the CH vertices contributes three skeletal electrons. Twelve of these skeletal electrons are used for the σ-bonding and the remaining six electrons for the π-bonding.

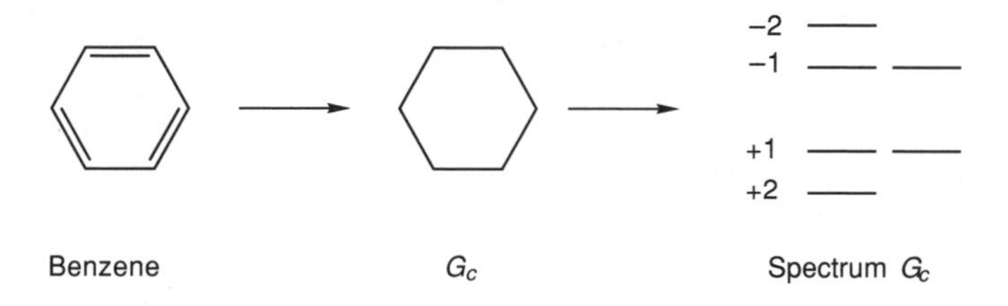

FIGURE 4–3. Benzene and the corresponding spectrum of G_C.

Figure 4–4 illustrates how the delocalized bonding in benzene from the C_6 overlap of the unique internal orbitals, namely the p orbitals, leads to aromatic stabilization. In a hypothetical localized "cyclohexatriene" structure, where the interactions between the p orbitals on each carbon atom are pairwise interactions, the corresponding graph G consists of three disconnected line segments (i.e., $3 \times K_2$). This graph has three +1 eigenvalues and three –1 eigenvalues. Filling each of the corresponding three bonding orbitals with an electron pair leads to an energy of 6β from this π bonding. In a delocalized "benzene" structure, where the delocalized interactions between the p orbitals on each carbon atom are described by the cyclic C_6 graph, filling the three bonding orbitals with an electron pair each leads to an energy of 8β. This corresponds to a resonance stabilization of $8\beta - 6\beta = 2\beta$, arising from the delocalized bonding of the carbon p orbitals in benzene. We will see below how similar ideas can be used to describe the three-dimensional aromaticity in deltahedral boranes.

–2

–1 –1

+1 +1

+2

Localized cyclohexatriene

Delocalized benzene

energy = [6(+1)]β = 6β

energy = [4(+1) + 2(+2)] =8β

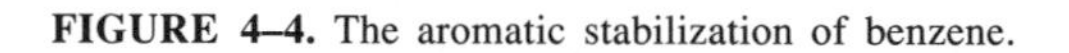

FIGURE 4–4. The aromatic stabilization of benzene.

[14]E. Hückel, *Z. Physik.*, **76**, 628, 1932.

Figure 4–5 illustrates the deltahedra found in the deltahedral boranes $B_nH_n^{2-}$. All of these deltahedra have only degree 4 and degree 5 vertices, except for one degree 6 vertex in the 11-vertex edge-coalesced icosahedron found, for example, in $B_{11}H_{11}^{2-}$. An 11-vertex deltahedron in which all of the vertices have degrees 4 or 5 is known to be topologically impossible.[15]

An important question is the nature of the core bonding graph G_c for the deltahedral boranes $B_nH_n^{2-}$. The two limiting possibilities for G_c are the complete graph K_n and the deltahedral graph D_n, and the corresponding core bonding topologies can be called the *complete* and *deltahedral* topologies, respectively. In the complete graph K_n, each vertex has an edge going to every other vertex leading to a total of $n(n-1)/2$ edges (Section 1.2). For any value of n, the complete graph K_n has only one positive eigenvalue, namely $n-1$, and $n-1$ negative eigenvalues, namely -1 each. The deltahedral graph D_n is identical to the 1-skeleton of the deltahedron depicted in Figure 4–6 for the corresponding value of n ($6 \leq n \leq 12$). Thus, two vertices of D_n are connected by an edge if and only if the corresponding vertices of the deltahedron are connected by an edge. The graphs D_n for the deltahedra of interest with six or more vertices (Figure 4–6) all have at least four zero or positive eigenvalues, designated as S^σ and P^σ orbitals in Figure 4–6. (The S^σ and P^σ terminology relates to tensor surface harmonic theory discussed in Section 4.5.) However, in all cases there is a unique positive eigenvalue which is much more positive than any other of the positive eigenvalues. This unique positive eigenvalue can be called conveniently the *principal eigenvalue* and corresponds to the fully symmetric $A_{(1)(g)}$ irreducible representation of the symmetry group of G_c. The molecular orbital corresponding to the principal eigenvalue of G_c may be called the *principal* core orbital. Since deltahedral boranes of the stoichiometry $B_nH_n^{2-}$ have $2n + 2$ skeletal electrons of which $2n$ are used for the surface bonding, as noted above, there are only two skeletal electrons available for core bonding, corresponding to a single core bonding molecular orbital and a single positive eigenvalue for G_c. Thus, deltahedral boranes are three-dimensional aromatic systems having $4k + 2 = 2$ core bonding electrons where $k = 0$, analogous to the $4k + 2$ π electrons where $k = 0$ ($C_3H_3^+$, $k = 1$ ($C_5H_5^-$, C_6H_6, $C_7H_7^+$) or $k = 2$ ($C_8H_8^{2-}$) for planar two-dimensional polygonal aromatic systems. Furthermore, only if G_c is taken to be the corresponding complete graph K_n will the simple model given above for globally delocalized deltahedra give the correct number of skeletal electrons in all cases, namely $2n + 2$ skeletal electrons for $6 \leq n \leq 12$. Such a model with complete core bonding topology can be used as a working basis for the chemical bonding topology of deltahedral boranes as well as the metal clusters discussed later in this book. However, deltahedral core bonding topology can also account for the observed $2n + 2$ skeletal electrons in the $B_nH_n^{2-}$

---

[15] R. B. King and A. J. W. Duijvestijn, *Inorg. Chim. Acta*, **178**, 55 , 1990.

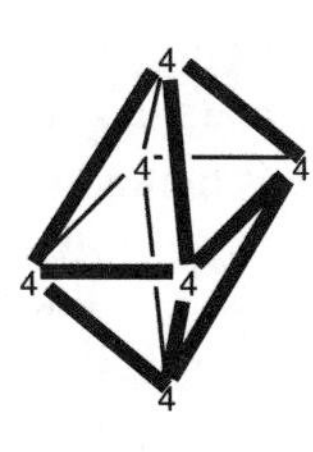

Octahedron

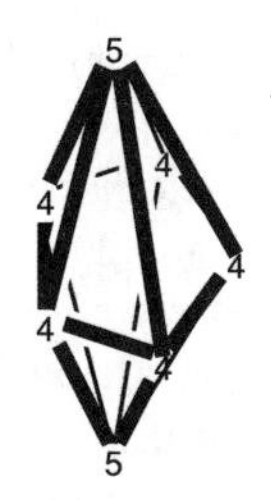

Pentagonal Bipyramid

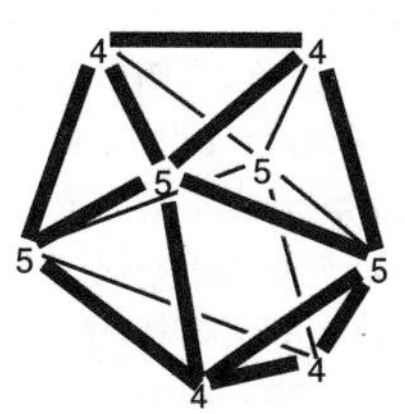

Bisdisphenoid
("D_{2d} Dodecahedron")

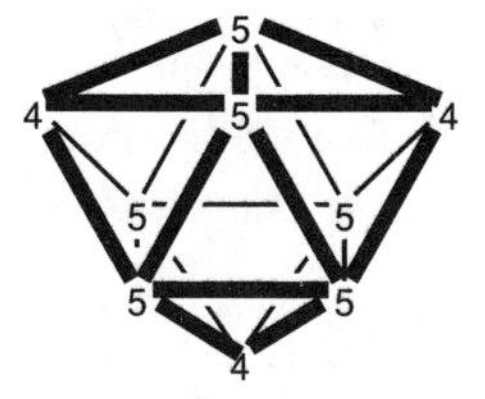

4,4,4-Tricapped Trigonal Prism

4,4-Bicapped Square Antiprism

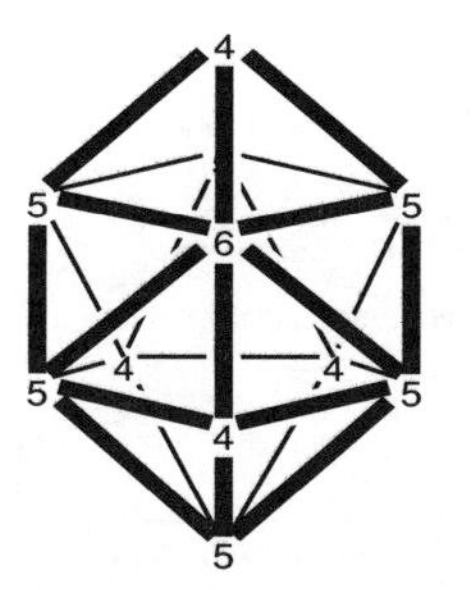

Edge-coalesced Icosahedron

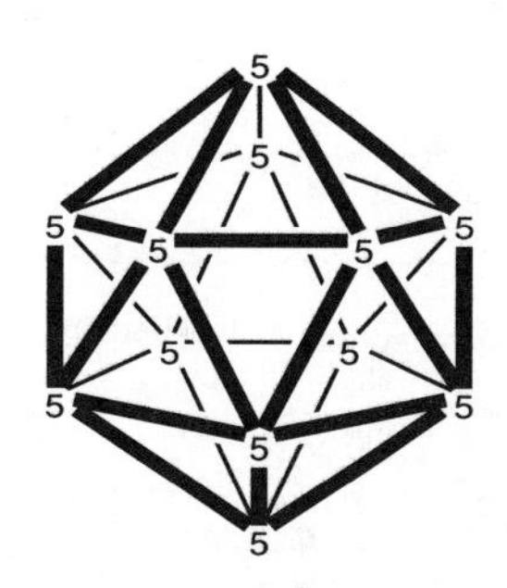

Regular Icosahedron

FIGURE 4–5. The deltahedra found in the deltahedral boranes $B_nH_n^{2-}$. The numbers in the vertices indicate their degrees. Note that all of the vertices have degrees 4 or 5 except for one degree 6 vertex of the 11-vertex edge-coalesced icosahedron.

deltahedral boranes, if there is a mechanism for raising the energies of all of the core molecular orbitals other than the principal core orbital to antibonding energy levels.

The distinction between complete (K_n) and deltahedral (D_n) core bonding topology is illustrated for octahedral $B_6H_6^{2-}$ in Figure 4–7. Among the (6)(6–1)/2 = 15 pairs of six vertices in an octahedron (D_6 graph), 12 pairs correspond to edges of the octahedron (*cis* interactions) and the remaining three pairs correspond to antipodal vertices related by the inversion center and *not* connected by an edge (*trans* interactions). However, all of the 15 pairs of six vertices in a complete K_6 graph correspond to edges of equal weight. In an octahedral array of six points, a parameter t can be defined as the ratio of the *trans* interactions to the *cis* interactions. This parameter t is 0 for the pure octahedral topology (D_6) and 1 for pure complete topology (K_6). Values of t between 0 and 1 can be used to measure gradations of topologies between D_6 and K_6, corresponding to the weighting of edges representing *trans* interactions relative to those representing *cis* interactions in the underlying graph G_c. In group-theoretical terms, pure complete core bonding topology (i.e., $t = 1$) uses the symmetric permutation group (see Section 2.5)[16] P_6 with 720 operations rather than its subgroup O_h with 48 operations (the symmetry point group of the octahedron) to describe the symmetry of the core bonding manifold in $B_6H_6^{2-}$. The actual O_h point group, rather than P_6 symmetry of $B_6H_6^{2-}$, results in the partial removal of the five-fold degeneracy of the core antibonding orbitals implied by the complete core bonding topology (Figure 4–7).

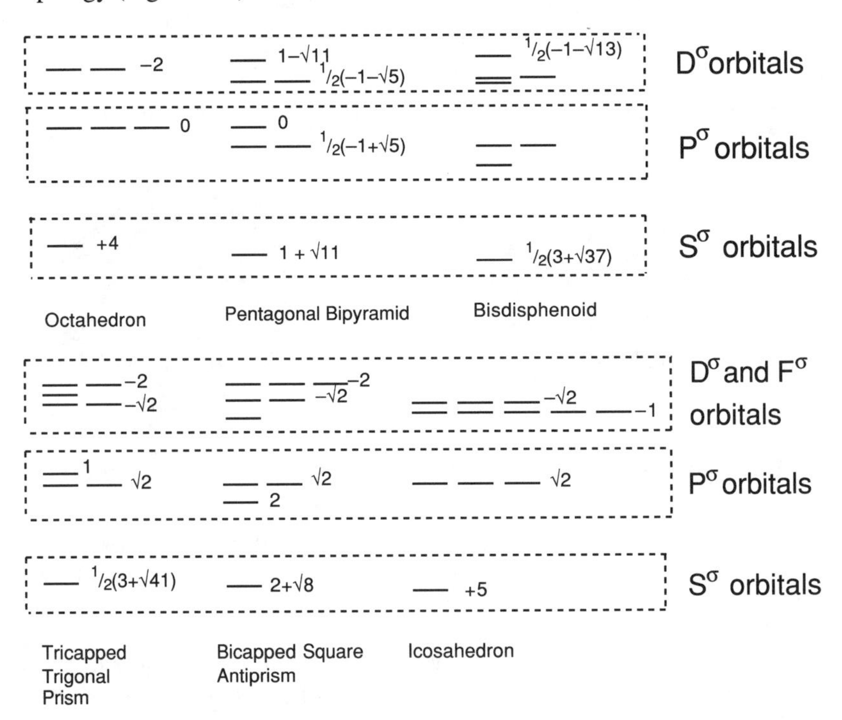

FIGURE 4–6. Spectra of the D_n graphs and the S^σ, P^σ, D^σ, and F^σ orbitals for the deltahedra with 6, 7, 8, 9, 10, and 12 vertices depicted in Figure 4-5.

---

[16]C. D. H. Chisholm, *Group Theoretical Techniques in Quantum Chemistry*, Academic Press, New York, 1976, chap. 6.

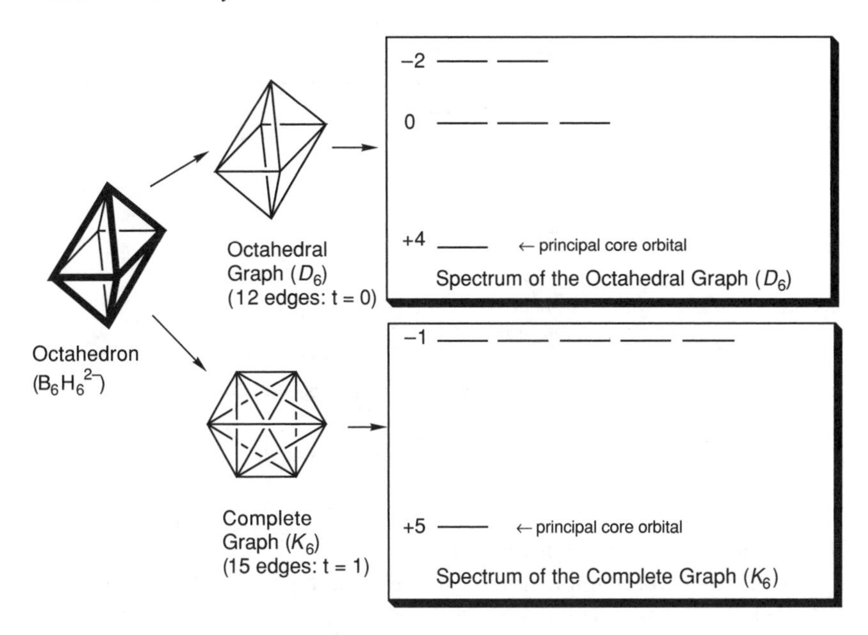

FIGURE 4–7. Spectra of the Octahedron (e.g., $B_6H_6^{2-}$) with the Deltahedral Topology (Octahedral Graph) and the Complete Topology.

Figure 4–8 shows how the delocalized bonding in $B_6H_6^{2-}$ arising from overlap of the unique internal orbitals, namely the radial *sp* hybrids on each boron atom, can lead to aromatic stabilization. In a hypothetical localized structure in which the interactions between the radial *sp* hybrids are pairwise interactions, the spectrum of the corresponding graph G_c is three disconnected line segments (i.e., $3 \times K_2$). The spectrum of this disconnected graph has three +1 eigenvalues and three –1 eigenvalues. Filling one of the bonding orbitals with the available two core bonding electrons leads to an energy of 2β from the core bonding. In a completely delocalized structure in which the core bonding is described by the complete graph K_6, this electron pair is in a bonding orbital with an eigenvalue of +5, corresponding to an energy of $(2)(5\beta) = 10\beta$ (Figure 4–8). The aromatic stabilization of completely delocalized $B_6H_6^{2-}$ is thus $10\beta - 2\beta = 8\beta$ assuming the same β unit for both the localized and complete delocalized structures. In an octahedrally delocalized $B_6H_6^{2-}$ in which the core bonding is described by the deltahedral graph D_6, corresponding to the 1-skeleton of the octahedron, the core bonding electron pair is in a bonding orbital with an eigenvalue of +4, corresponding to an energy of $(2)(4\beta) = 8\beta$ (Figure 4–8). The aromatic stabilization of octahedrally delocalized $B_6H_6^{2-}$ is thus $8\beta - 2\beta = 6\beta$. Thus the aromatic stabilization of $B_6H_6^{2-}$ is considerable, regardless of whether the delocalized core bonding is onsidered to have the complete topology represented by the complete graph K_6 or the octahedral topology represented by the deltahedral graph D_6.

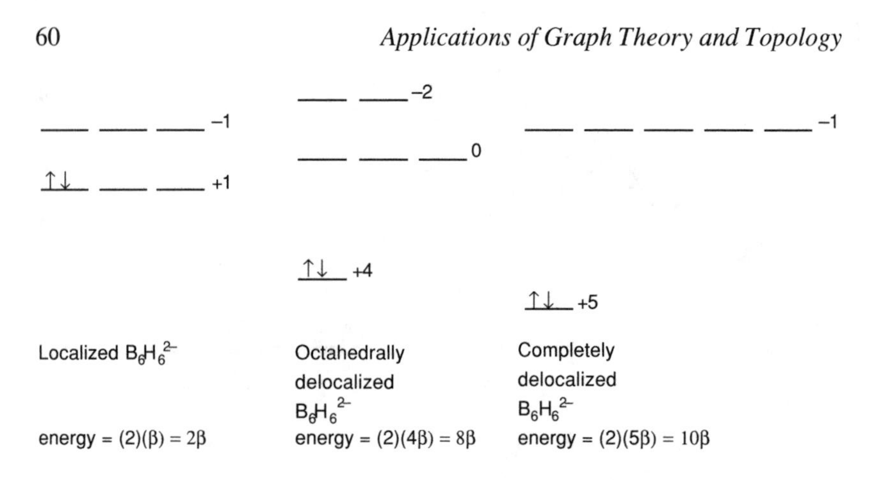

FIGURE 4–8. The aromatic stabilization of $B_6H_6^{2-}$.

The essential features of the bonding in globally delocalized polygonal and deltahedral clusters are summarized in Table 4–2.

There are several implications of the bonding model for delocalized deltahedral structures with n vertices using complete core bonding topology described by the corresponding K_n complete graph:

TABLE 4–2
Skeletal Bonding in Globally Delocalized Polygons and Polyhedra

	Polygonal Structures	Deltahedral Structures
(A) Dimensionality	2	3
(B) Orbital Hybridizations		
External	sp^2	sp
Twin Internal	sp^2	p
Unique Internal	p	sp
(C) Orbital Interactions		
(1) Twin Internal		
(a) Bonding Type	σ (peripheral)	surface
(b) Interaction Topology	nK_2	nK_2
(c) Number of Bonding Orbitals	n	n
(d) Number of Bonding Electrons	$2n$	$2n$
(2) Unique Internal		
(a) Nodality	1	0
(b) Bonding Type	π	core
(c) Interaction Topology	C_n	K_n
(d) Number of Bonding Orbitals	odd $(2k+1)$	1
(e) Number of Bonding Electrons	$4k+2$	2

1. The overlap of the n unique internal orbitals to form an n-center core bond may be hard to visualize since its topology corresponds to that of the complete graph K_n which for $n \geq 5$ is non-planar by Kuratowski's theorem (Section 1.2) and thus cannot correspond to the 1-skeleton of a polyhedron realizable in three-dimensional space. However, the overlap of these unique internal orbitals does not occur along the edges of the deltahedron or along those of any other three-dimensional polyhedron. For this reason, the topology of the overlap of the unique internal orbitals in the core bonding of a deltahedral cluster need not correspond to a graph representing a 1-skeleton of a three-dimensional polyhedron. The only implication of the K_n graph description of the bonding topology of the unique internal orbitals is that the deltahedron is topologically homeomorphic to the sphere.
2. The equality of the interactions between all possible pairs of unique internal orbitals required by the K_n model for the core bonding is obviously a very crude assumption, since in any deltahedron with five or more vertices, all pairwise relationships of the vertices are not equivalent. The example of the non-equivalence of the *cis* and *trans* vertex pairs in an octahedral structure such as $B_6H_6^{2-}$ has already been discussed. However, the single eigenvalue of the K_n graph is so strongly positive that severe inequalities in the different vertex pair relationships are required before the spectrum of the graph representing precisely the unique internal overlap contains more than one positive eigenvalue.

4.4 ELECTRON RICH POLYHEDRAL SYSTEMS

Electron-rich polyhedral systems are those containing more than the $2n + 2$ skeletal electrons required for globally delocalized n-vertex deltahedra without vertices of degree 3. In the case of boron hydride derivatives,[17,18] there are well-known families of *nido* compounds having $2n + 4$ skeletal electrons and *arachno* compounds having $2n + 6$ skeletal electrons. In the *nido* polyhedra, all but one of the faces are triangular; the unique non-triangular face may be regarded as a hole. Analogously, the *arachno* polyhedra have either two non-triangular faces or one large non-triangular face (i.e., two holes or one large bent hole). Thus, successive additions of electron pairs to a closed $2n + 2$ skeletal electron deltahedron result in successive punctures of the deltahedral surface to give holes (faces) having more than three edges by a process conveniently called *polyhedral puncture*. The open polyhedral networks can also be considered to arise by excision of one or more vertices along with all of the edges leading to them from a closed deltahedron having

[17]E. L. Muetterties and W. H. Knoth, *Polyhedral Boranes*, Marcel Dekker, New York, 1968.
[18]E. L. Muetterties, Ed., *Boron Hydride Chemistry*, Academic Press, New York, 1975.

$m > n$ vertices by a process conveniently called *polyhedral excision*. Figure 4–9 shows the *nido* and *arachno* polyhedra (and polyhedral fragments) derived from the octahedron, pentagonal bipyramid, and regular icosahedron.

Let us consider the treatment of the skeletal bonding topology in *nido* polyhedra, namely polyhedra with only one non-triangular face and $2n + 4$ skeletal electrons. In the treatment of such structures, the vertex atoms may be divided into the following two sets: border vertex atoms, which are vertices of the one face containing more than three edges (i.e., they are at the border of the single hole), and interior vertex atoms, which form vertices of only triangular faces. For example, in a square pyramid (Figure 4–9), which is the simplest example of a *nido* polyhedron, the four basal vertices are the border vertices, since they all border the square "hole", i.e., the base of the square pyramid. However, the single apical vertex of the square pyramid is an interior vertex, since it is a vertex of only triangular faces. The external and twin internal orbitals of the border vertex atoms are taken to be sp^2 hybrids. The unique internal orbitals of the border vertex atoms will thus be p orbitals. The external and unique internal orbitals of the interior vertex atoms are taken to be sp hybrids, in accord with the treatment of closed deltahedra discussed in the previous section. The twin internal orbitals of the interior vertex atoms must therefore be p orbitals. Note that in the *nido* polyhedra, the hybridization of the border vertex atoms is the same as that of the vertex atoms of polygonal systems, whereas the hybridization of the interior vertex atoms is the same as that of the vertex atoms of deltahedral systems. A chemical consequence of the similar vertex atom hybridizations in polygons and the borders of *nido* polyhedra is the ability of both planar polygonal hydrocarbons (e.g., cyclopentadienyl and benzene) and the border atoms of *nido* carboranes[19] to form chemical bonds with transition metals of similar types, involving interaction of the transition metal with *all* of the atoms of the planar polygon or the border atoms of the polygonal hole of the *nido* polyhedron.

Nido polyhedra can be classified into two fundamental types: the pyramids with only one interior vertex (the apex) and the non-pyramids with more than one interior vertex. If n is the total number of vertices and v is the number of interior vertices in a *non-pyramidal nido* polyhedron, the interactions between the internal orbitals which generate bonding orbitals are of the following three different types:

1. The $2(n - v)$ twin internal orbitals of the border atoms and the $2v$ twin internal orbitals of the interior atoms interact along the polyhedral surface to form n bonding orbitals and n antibonding orbitals;

[19]M. F. Hawthorne, *Acc. Chem. Res.*, **1**, 281, 1968.

$2n + 2$ electrons
6 vertices
octahedron
$C_2B_4H_6$
"closo"

$2n + 4$ electrons
5 vertices
square pyramid
B_5H_9
"nido"

$2n + 6$ electrons
4 vertices
butterfly
B_4H_{10}
"arachno"

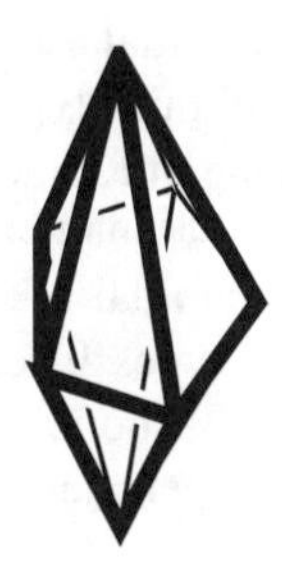

$2n + 2$ electrons
7 vertices
pentagonal bipyramid
$C_2B_5H_7$
"closo"

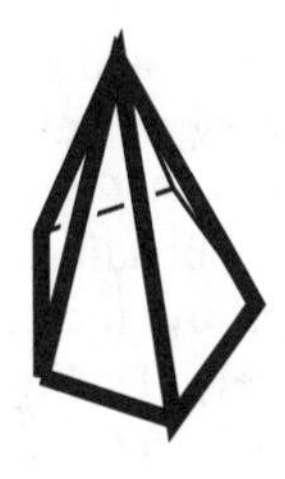

$2n + 4$ electrons
6 vertices
pentagonal pyramid
B_6H_{10}
"nido"

$2n + 6$ electrons
5 vertices

B_5H_{11}
"arachno"

$2n + 2$ electrons
12 vertices
icosahedron
$C_2B_{10}H_{12}$
"closo"

$2n + 4$ electrons
11 vertices
$C_2B_9H_{11}^{2-}$
"nido"

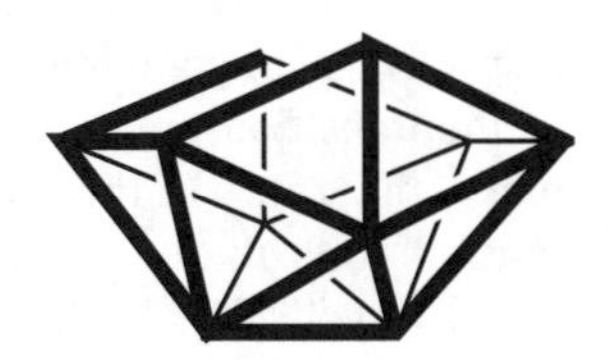

$2n + 6$ electrons
10 vertices
$B_{10}H_{14}^{2-}$
"arachno"

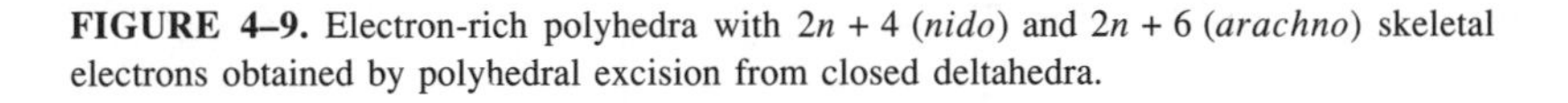

FIGURE 4–9. Electron-rich polyhedra with $2n + 4$ (*nido*) and $2n + 6$ (*arachno*) skeletal electrons obtained by polyhedral excision from closed deltahedra.

2. The v unique internal orbitals of the interior vertex atoms all interact with each other at the core of the structure in a way which may be represented by the complete graph K_v to give a single bonding orbital and $v - 1$ antibonding orbitals;
3. The $n - v$ unique internal orbitals of the border atoms interact with each other across the surface of the hole in a way which may be represented by the complete graph K_{n-v} to give a single bonding orbital and $n - v - 1$ antibonding orbitals.

The above interactions in *nido* systems of the first two types correspond to the interactions found in the closed deltahedral systems discussed in the previous section, whereas the interactions of the third type can only occur in polyhedra containing at least one hole, such as the non-pyramidal *nido* systems. Furthermore, the values of v and $n–v$ in the second and third types of interactions are immaterial as long as they both are greater than one, since any complete graph K_i ($i > 1$) has exactly one positive eigenvalue, namely $i - 1$. The total number of skeletal bonding orbitals in non-pyramidal *nido* systems with n vertices generated by interactions of the three types listed above are n, 1, and 1, respectively, leading to a total of $n + 2$ bonding orbitals holding $2n + 4$ skeletal electrons in accord with experimental observations.

Pyramidal *nido* polyhedra having only one interior vertex require a somewhat different treatment because the eigenvalue of the one-vertex no-edge complete graph K_1 is zero, leading to ambiguous results for the second type of interaction listed above. This difficulty can be circumvented by realizing that the only types of pyramids relevant to delocalized borane and metal cluster chemistry are square pyramids, pentagonal pyramids, and hexagonal pyramids, and bonding schemes for these types of pyramids can be constructed which are completely analogous to well-known[20] transition metal complexes of cyclobutadiene, cyclopentadienyl, and benzene, respectively. In applying this analogy, the interior vertex atom plays the role of the transition metal, and the planar polygon of the border vertex atoms plays the role of the planar polygonal ring in the metal complexes. Furthermore, the $n - 1$ unique internal orbitals of the border vertex atoms interact cyclically, leading to three "submolecular" orbitals, which may be used for bonding to the single interior vertex atom as represented by the three non-negative eigenvalues of the corresponding C_{n-1} cyclic graph ($n = 5, 6, 7$). Of these three polygonal orbitals, one orbital, the A_1 orbital, has no nodes perpendicular to the polygonal plane, whereas the other two remaining orbitals, the degenerate E orbitals, each have one node perpendicular to the polygonal plane (Figure 4–10). These two nodes from the degenerate E orbitals are mutually perpendicular, as depicted in Figure 4–10.

---

[20]R. B. King, *Transition Metal Organometallic Chemistry: An Introduction*, Academic Press, New York, 1969.

The following three interactions are used to generate the skeletal bonding orbitals in *nido* pyramids:

1. The $2(n - 1)$ twin internal orbitals of the border atoms interact along the edges of the base of the pyramid to form $n - 1$ bonding orbitals and $n - 1$ antibonding orbitals analogous to the σ bonding and σ^* antibonding orbitals, respectively, of planar polygonal hydrocarbons.
2. The unique internal orbital of the single interior vertex atom (the apex of the pyramid) interacts with the A_1 orbital to give one bonding orbital and one antibonding orbital.

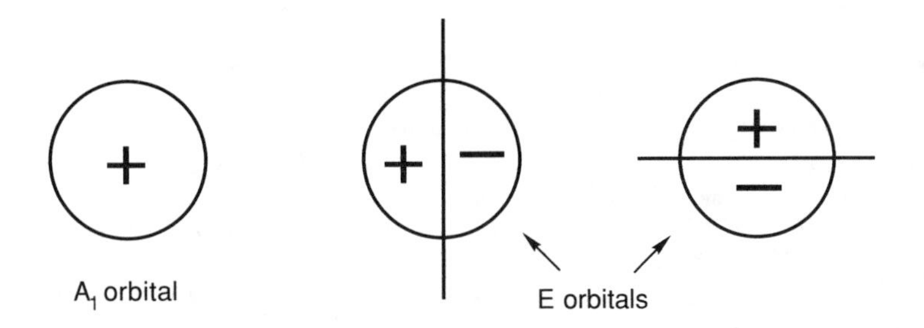

FIGURE 4–10. The "submolecular" orbitals arising from interaction of the n–1 unique internal orbitals of the vertex atoms bordering the non-triangular face in the *nido* polyhedron.

3. The twin internal orbitals of the apex of the pyramid interact with the two orthogonal E orbitals in two separate pairwise interactions to give two bonding and two antibonding orbitals.

The total number of skeletal bonding orbitals in pyramidal *nido* systems generated by these three interactions are n–1, 1, and 2, respectively, leading to a total of $n + 2$ bonding orbitals holding $2n + 4$ skeletal electrons. Thus the graph-theoretical treatment of non-pyramidal and pyramidal *nido* polyhedra with n vertices leads to the prediction of the same numbers of skeletal bonding orbitals, namely $n + 2$, in accord with experimental observations. However, the partitionings of these bonding orbitals are different for the two types of *nido* systems, namely (n, 1, 1) for the non-pyramidal systems and (n–1, 1, 2) for the pyramidal systems.

The key features of this skeletal bonding model for *nido* polyhedra are summarized in Table 4–3.

The process of polyhedral puncture which forms *nido* polyhedra with one hole and $2n + 4$ skeletal electrons from closed deltahedra with $2n + 2$ skeletal electrons can be continued further to give polyhedral fragments containing two or more holes or one larger hole. In boron chemistry, polyhedra or polyhedral fragments having $2n + 6$ and $2n + 8$ skeletal electrons are called *arachno* and *hypho* structures, respectively. Some examples of *arachno* structures are shown in Figure 4–9. Formation of a new hole by such

TABLE 4–3
Skeletal Bonding in *Nido* Polyhedra

	Non-Pyramids		Pyramids	
	Border	Interior	Base	Apex
(A) Orbital Hybridizations				
External	sp^2	sp	sp^2	sp
Twin Internal	sp^2	p	sp^2	p
Unique Internal	p	sp	p	sp
(B) Orbital Interactions				
(1) Twin Internal Orbitals				
(a) Bonding Type	surface	surface	σ	surface
(b) Interaction Topology	$(n-v)K_2$	vK_2	$(n-1)K_2$	$2K_2$
(c) Number of Bonding Orbitals	$n-v$	v	$n-1$	2
(d) Number of Bonding Electrons	$2(n-v)$	$2v$	$2(n-1)$	4
(2) Unique Internal Orbitals				
(a) Bonding Type	hole	core	*	core
(b) Interaction Topology	K_n	K_n	*	K_2
(c) Number of Bonding Orbitals	1	1	0*	1
(d) Number of Bonding Electrons	2	2	0*	2

* The unique internal orbitals of the base (i.e, border) atoms of the pyramidal systems interact according to the C_n topology (n = 4,5,6) similar to the π-bonding scheme in the polygonal systems. The resulting A_1 and E molecular orbitals then interact further with the unique internal and twin internal orbitals, respectively, of the apex atom to form the bonding orbitals listed in this table. However, since the bonding orbitals from the interaction of the unique internal orbitals of the base atom of pyramidal systems are used further for these interactions with the apex atom internal orbitals, this interaction is not listed explicitly in this table. See text for further details.

polyhedral puncture splits the complete graph formed by interactions at the polyhedral core between the unique internal orbitals of the interior vertex atoms into two smaller complete graphs.

One of these new complete graphs involves interaction at the polyhedral core between the unique internal orbitals of the vertex atoms which are still interior atoms after creation of the new hole or expansion of the existing hole. The second new complete graph involves interaction above the newly created hole between the unique internal orbitals of the vertex atoms which have become border atoms of the newly created hole. Since each new complete graph contributes exactly one new skeletal bonding orbital to the polyhedral system, each application of polyhedral puncture to give a stable system requires addition of two skeletal electrons.

4.5 TENSOR SURFACE HARMONIC THEORY

The graph-theory derived model for the skeletal bonding of a globally delocalized deltahedral borane with n vertices with complete core bonding topology outlined in Section 4.3 uses the corresponding complete graph K_n to describe the topology of the multicenter core bond. The precise topology of the cluster deltahedron does not enter directly into such models; only the absence of degree 3 vertices enters in. In other words, graph-theory derived models of the skeletal bonding of globally delocalized deltahedral clusters consider such deltahedra to be topologically homeomorphic to the sphere.[21]

The topological homeomorphism of a deltahedron to a sphere used in the graph-theory derived models is also the basis of the tensor surface harmonic theory developed by Stone.[22,23,24,25] The tensor surface harmonic theory defines the vertices of a deltahedral borane as lying on the surface of a single sphere with the atom positions described by the standard angular coordinates θ and ϕ (Equations 3–2) related to latitude and longitude. The second order differential equations for the angular dependence of the molecular orbitals from the core bonding become identical to the equations for the angular dependence of the atomic orbitals obtained by solution of the Schrödinger equation, with both sets of equations making use of the spherical harmonics $Y_{LM}(\theta,\phi)$ discussed in Section 3.1.

In tensor surface harmonic theory as applied to deltahedral boranes, the internal orbitals of the vertex atoms are classified by the number of nodes with respect to the radial vector connecting the vertex atom with the center of the deltahedron.[26] The unique internal orbitals (namely *sp* hybrids, as indicated in Table 4–2) are nodeless (i.e., σ-type as depicted in Figure 4–2) and lead to core bonding molecular orbitals described by the scalar spherical harmonics $\Theta(\theta)\cdot\Phi(\phi) = Y_{LM}(\theta,\phi)$, which, for deltahedra having n vertices, correspond successively to a single nodeless S^σ orbital (Y_{00}) the three uninodal P^σ orbitals (Y_{10}, Y_{11c}, Y_{11s}), the five binodal D^σ orbitals ($Y_{20}, Y_{21c}, Y_{21s}, Y_{22c}, Y_{22s}$), the seven trinodal F^σ orbitals ($Y_{30}, Y_{31c}, Y_{31s}, Y_{32c}, Y_{32s}, Y_{33c}, Y_{33s}$), etc., of increasing energy. The S^σ, P^σ, D^σ, F^σ orbitals, etc., correspond to the molecular orbitals arising from the n-center core bond of the deltahedron. The energy levels of these orbitals for the core bonding in six of the seven deltahedra depicted in Figure 4–5 correspond to the spectra of the corresponding deltahedral graphs D_n and are shown in Figure 4–6 with enough

---

[21] N. J. Mansfield, *Introduction to Topology*, Van Nostrand, Princeton, New Jersey, 1963, 40.

[22] A. J. Stone, *Mol. Phys.*, **41**, 1339, 1980.

[23] A. J. Stone, *Inorg. Chem.*, **20**, 563, 1981.

[24] A. J. Stone and M. J. Alderton, *Inorg. Chem.*, **21**, 2297, 1982.

[25] A. J. Stone, *Polyhedron*, **3**, 1299, 1984.

[26] R. L. Johnston and D. M. P. Mingos, *Theor. Chim. Acta*, **75**, 11, 1989.

of the eigenvalues indicated in the diagrams to clearly define the scales. Note that the S^σ and P^σ molecular orbitals appear in well-separated groups, whereas the clearly antibonding D^σ and F^σ molecular orbitals appear clustered around eigenvalues of –1 to –2 without a clear separation.

The twin internal orbitals (namely *p* orbitals as indicated in Table 4–2) are uninodal (i.e., π-type as depicted in Figure 4–2) and lead to surface bonding described by the vector surface harmonics. Two vector surface harmonic functions can be generated from each Y_{LM} as follows:

$$V_{LM} = \nabla Y_{LM} \tag{4–14a}$$

$$\overline{V}_{LM} = \mathbf{r} \times \nabla Y_{LM} \tag{4–14b}$$

In Equations 4–14 ∇ is the vector operator,

$$\nabla = \left(\frac{\partial}{\partial\theta}, \frac{1}{\sin\theta}\frac{\partial}{\partial\phi} \right) \tag{4–15}$$

$\times$ is the vector cross-product, and the $\overline{V}_{LM}$ of Equation 4–14b is the parity inverse of the V_{LM} of equation 4–14a, corresponding to a rotation of each atomic π-function by 90° about the radial vector **r**. The V_{LM} and $\overline{V}_{LM}$ correspond to the equal numbers of bonding and antibonding surface orbitals in a globally delocalized deltahedral cluster leading to three P^π, five D^π, seven F^π, etc., bonding/antibonding orbital pairs of increasing energy and nodality. Since Y_{00} is a constant, $\nabla Y_{00} = 0$ so that there are no S^π or $\overline{S}^\pi$ orbitals.

The core and surface orbitals defined above by tensor surface harmonic theory can be related to the following aspects of the graph-theory derived model for the skeletal bonding in boranes with the deltahedral structures depicted in Figure 4–6:

1. The lowest energy fully symmetric core orbital (A_{1g}, A_g, A_1, or A_1' depending upon the point group of the deltahedron) corresponds to the S^σ orbital in tensor surface harmonic theory. Since there are no S^π or $\overline{S}^\pi$ surface orbitals, this lowest energy core orbital cannot mix with any surface orbitals, so that it cannot become antibonding through core-surface mixing.
2. The three core orbitals of next lowest energy correspond to P^σ orbitals in tensor surface harmonic theory. These orbitals can mix with the P^π surface orbitals so that the P^σ core orbitals become antibonding with corresponding lowering of the bonding energies of the P^π surface orbitals below the energies of the other surface orbitals. This is why graph-theory derived models of skeletal bonding in globally delocalized *n*-vertex deltahedra, which use the K_n graph to describe the multicenter

core bond, give the correct numbers of skeletal bonding orbitals even for deltahedra whose corresponding deltahedral graph D_n has more than one positive eigenvalue. In this way, tensor surface harmonic theory can be used to justify important assumptions in the graph-theory derived models.

Chapter 5

RELATIONSHIP OF TOPOLOGICAL TO COMPUTATIONAL METHODS FOR THE STUDY OF DELOCALIZED BORANES

5.1 TOPOLOGICAL ANALYSIS OF COMPUTATIONAL RESULTS

The previous chapter summarized a qualitative topological method based on graph theory for the study of delocalization in polyhedral boranes as well as planar polygonal aromatic hydrocarbons. This chapter compares this qualitative topological method for deltahedral boranes with various quantitative and semiquantitative computations on polyhedral boranes including both semi-empirical Hückel methods as well as *ab initio* methods. Such comparisons are useful not only to justify the applicability of the qualitative methods but also to interpret the numerical results of the computations.

The molecular orbital parameters for a delocalized system such as the deltahedral boranes can be related to the eigenvalues of the adjacency matrix of the underlying graph G by the following equation (≡ Equation 4–12):

$$E_k = \frac{\alpha + x_k\beta}{1 + x_kS} \tag{5–1}$$

In this equation α is the standard Coulomb integral, β is the resonance integral, and S is the overlap integral between atomic orbitals on neighboring atoms (Section 4.1). In order to use Equation 5–1 to interpret a set of computed molecular orbital energy parameters, the three parameters α, β, and S must be determined. However, any actual system provides too few relationships to determine fully all of these three parameters. Therefore some assumptions concerning the values of α, β, and S are necessary for any comparisons to be feasible. The generally used approach (discussed in Section 4.1) first assumes a zero value for S and then determines α from the midpoint of all of the molecular energies by taking an appropriately weighted average. The third parameter, β, can then be determined from specific orbital energies.

In order to relate a given computation on a deltahedral borane to topological models for its chemical bonding, computed values for all of the molecular orbital energy parameters are required, including those for the unfilled antibonding (virtual) orbitals.[1] Having such information, the first step is to calculate α, the energy "zero point" in Equation 5–1. In the case of

[1]R. B. King, B. Dai, and B. M. Gimarc, *Inorg. Chim. Acta*, **167**, 213, 1990.

a deltahedral borane anion $B_nH_n^{2-}$, this can be done by taking the mean of the energy parameters for all $5n$ molecular orbitals arising from the $4n$ atomic orbitals of the n sp^3 boron manifolds and the n s orbitals of the n hydrogen atoms. In taking this mean, the degenerate orbitals of the E, T, G, and H representations are given the weights 2, 3, 4, and 5, respectively, corresponding to their degeneracies. Hence the parameter α becomes

$$\alpha = \frac{\sum_k g_k E_k}{\sum_k g_k} \qquad (5\text{–}2)$$

In Equation 5–2, g_k is the degeneracy of energy level k, and the summation is over all molecular orbitals k. It is convenient to tabulate orbital energies as $E_k' = E_k - \alpha$, such that

$$\sum_k E_k' = 0 \qquad (5\text{–}3)$$

The surface energy unit, β_s, can also be estimated at this stage as the degeneracy-weighted average distance of the pure surface orbitals from the energy zero point α. At this stage, an unavoidable sampling error is introduced, since only the pure surface orbitals can be included in the averages. The energy parameters of the other surface orbitals must be excluded from this average since they are distorted by substantial core-surface and external-surface mixing. This point will be clarified later in the actual examples.

Further analysis of the computed energy parameters either requires some special symmetry, such as that found in octahedral $B_6H_6^{2-}$ or icosahedral $B_{12}H_{12}^{2-}$, or some further assumptions concerning the chemical bonding topology for less symmetrical systems. In the cases of $B_6H_6^{2-}$ and $B_{12}H_{12}^{2-}$, the core energy units β_c and the non-adjacent atom unique internal orbital interactions can be estimated from the energy parameters of the principal core orbital and a second core orbital, which do not mix with the surface orbitals. Possible errors arising from core-external orbital interactions do not appear to be large.

In order to apply this method, it is necessary to know the irreducible representations corresponding to Γ_σ for the core and external orbitals and Γ_π for the surface orbitals for the deltahedra of interest. The group-theoretical procedure for calculating Γ_σ is given in Chapter 3; a similar procedure considering the effects of all of the symmetry operations of the point group G on a uninodal π orbital (e.g., Figure 4–2) located at each deltahedral vertex can be used to calculate Γ_π.[2] Table 5–1 gives the representations for Γ_σ and Γ_π for the deltahedra of interest in this chapter (Figure 4–5) in terms of the component irreducible representations. The pure surface orbitals are starred in Table 5–1; these orbitals correspond to irreducible representations found in Γ_π but not in Γ_σ.

TABLE 5–1
Orbital Representations for Deltahedra

Deltahedron	Γ_σ (core and external orbitals)	Γ_π (surface orbitals)
Octahedron	$A1_g+T_{1u}+E_g$	$T_{1u}+T_{2g}*+T_{2u}*+T_{1g}*$
Pentagonal bipyramid	$2A_1'+E_1'+E_2'+A_2''$	$A_2'*+2E_1'+E_2'+A_2''+2E_1''*+E_2''*$
Bisdisphenoid	$2A_1+2B_2+2E$	$2A_1+2A_2*+2B_1*+2B_2+4E$
4,4,4-Tricapped trigonal prism	$2A_1'+2E'+A_2''+E''$	$A_1'+2A_2'*+3E'+A_1''*+2A_2''+3E''$
4,4 Bicapped square antiprism	$2A_1+2B_2+E_1+E_2+E_3$	$A_2+A_2*+B_1*+B_2+3E_1+2E_2+3E_3$
Icosahedron	$A_g+T_{1u}+T_{2u}+H_g$	$T_{1u}+H_g+G_u*+G_g*+H_u*+T_{1g}*$

.C1.5.2 APPLICATIONS TO OCTAHEDRAL AND ICOSAHEDRAL BORANES;

The following computations on $B_6H_6^{2-}$ and $B_{12}H_{12}^{2-}$ have been analyzed by this method,[3] and the results for octahedral $B_6H_6^{2-}$ are summarized in Table 5–2:

1. Early extended Hückel computations by Hoffmann and Lipscomb[4] using a Slater orbital basis set (HL5*n* in Table 5–2);
2. Self-consistent molecular orbital computations by Armstrong, Perkins, and Stewart[5] also using a Slater orbital basis set (APS in Table 5–2);
3. *Ab initio* self-consistent field molecular orbital computations by Gimarc and collaborators[1] using a Gaussian 82 program with a STO-3G basis set (GD in Table 5–2).

---

[2]F. A. Cotton, *Chemical Applications of Group Theory*, John Wiley & Sons, New York, 1971.

[3]R. B. King, *J. Math. Chem.*, **4**, 69, 1990.

[4]R. Hoffmann and W. N. Lipscomb, *J. Chem. Phys.*, **36**, 2179, 1962.

[5]D. R. Armstrong, P. G. Perkins, and J. J. Stewart, *J. Chem. Soc. (A)*, 3674, 1971.

TABLE 5–2
Analysis of Computations on Octahedral $B_6H_6^{2-}$

	HL5*n*	**APS**	**GD**
Core Orbitals			
A_{1g} (principal)	+3.210	–50.3	–1.126
E_g	–0.888	+13.6	+0.470
T_{1u} (actual)	–0.844	11.3	–0.848
T_{1u} (effect of mixing removed)	(–0.478)	(10.1)	
Pure Surface Orbitals			
T_{2g}	+0.493	–5.5	–0.486
T_{2u}	–0.416	+9.8	+0.198
T_{1g}	–0.671	+11.7	+0.548
Derived Parameters			
α	0	+7.2	+0.675
β_s	0.527	–8.1	–0.429
β_c	0.683	–10.7	–0.266
β_c/β_s	1.296	1.320	0.620
t	0.700	0.700	0.233
$\Delta E(T_{1u})$	0.366	–1.2	

In order to study the data for octahedral $B_6H_6^{2-}$, consider an octahedrally weighted K_6 graph (Figure 5–1), having 12 edges of unit weight corresponding to the octahedron edges (*cis* interactions) and the remaining 3 edges of weight t corresponding to the three octahedron antipodal vertex pairs (*trans* interactions).[6] The spectrum of this graph has a non-degenerate eigenvalue $4 + t$ corresponding to the A_{1g} principal core orbital, a triply degenerate eigenvalue $-t$ corresponding to the triply degenerate T_{1u} core molecular orbital, and a doubly degenerate $-2 + t$ eigenvalue corresponding to the doubly degenerate E_g core molecular orbital (Figure 5–1). Note that any positive value of t (up to +2) is sufficient to lead to only one positive eigenvalue, namely the $4 + t$ eigenvalue of the A_g orbital, and five negative eigenvalues, namely the $-t$ eigenvalues of the triply degenerate T_{1u} orbital and the $-2 + t$ eigenvalues of the doubly degenerate E_g orbital. This indicates that any reasonable positive *trans* interaction in an octahedron gives the same distribution of bonding and antibonding orbitals, namely 1 and 5, respectively, as an unweighted ($t = 1$) K_6 graph. Thus, for octahedral boranes, the numbers of bonding and antibonding orbitals are insensitive to the value taken for t. Note also that setting $t = 0$ leads to the spectrum of the octahedron (+4,0,0,0,–2,–2), which is the D_6 graph, whereas setting $t = 1$ leads to the spectrum of the K_6 complete graph (+5,–1,–1,–1,–1,–1).

[6]R. B. King, *J. Comput. Chem.*, **8**, 341, 1987.

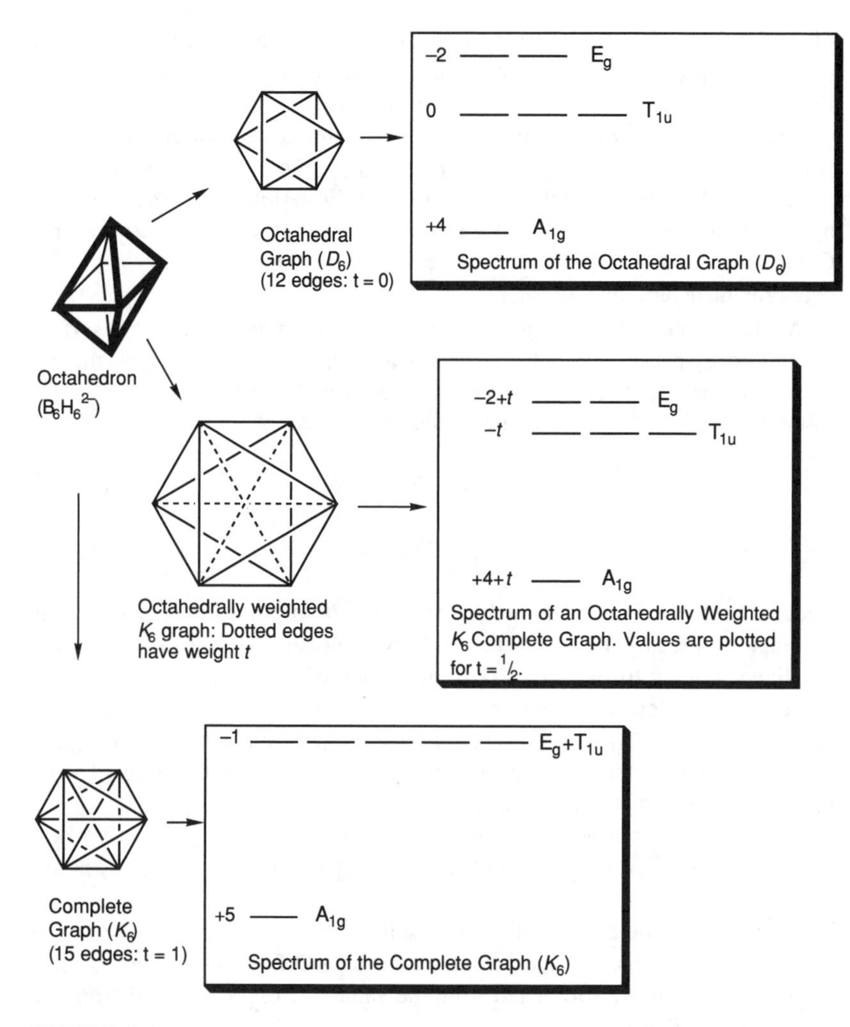

FIGURE 5–1. The spectrum of an octahedrally weighted K_6 complete graph as compared with that of the octahedron and the K_6 complete graph.

The spectrum of the octahedrally weighted K_6 complete graph indicates that in the absence of core-surface mixing, equation 4–13 for the energy parameters of the octahedral core orbitals in $B_6H_6^{2-}$ relative to α becomes the equations

$$E(A_{1g}) = (4 + t)\beta_c \tag{5–4a}$$

$$E(T_{1u})_c = -t\beta_c \tag{5–4b}$$

$$E(E_g) = (-2 + t)\beta_c \tag{5–4c}$$

Since the A_{1g} and E_g orbitals are pure core orbitals (i.e., there are no surface orbitals having these irreducible representations—see Table 5–1), the two Equations 5–4a and 5–4c can be used to calculate the two parameters t and β_c corresponding to the $E(A_{1g})$ bonding and $E(E_g)$ antibonding energy parameters from a given computation, provided $\alpha = 0$ or is already known. Substitution of these calculated values for t and β_c in Equation 5–4b then gives a hypothetical value for $E(T_{1u})_c$ in the absence of core-surface mixing. Comparison of this hypothetical value with the computed value for the T_{1u} core orbitals determines $\Delta E(T_{1u})$.

A related approach can be used to compare the computed octahedral surface orbital energy parameters with the ideal values arising from the graph-theory-derived method. In this case, the ideal surface orbital energy parameters are the following, with β_s designating the surface orbital energy unit:

$$E(T_{2g}) = E(T_{1u})_s = \beta_s \qquad (5\text{–}5a)$$

$$E(T_{2u}) = E(T_{1g}) = -\beta_s \qquad (5\text{–}5b)$$

Reduction of the effective symmetry from the P_6 automorphism group of the K_6 graph implied by the graph-theory-derived model using the complete core bonding topology to the actual O_h point group of a regular octahedron will make $E(T_{2u})$ no longer equal to $E(T_{1g})$ and $E(T_{2g})$ no longer equal to $E(T_{1u})_s$. On the basis of Equations 5–5a and 5–5b, the following appropriately weighted mean of the *pure* surface orbitals T_{2g}, T_{2u}, and T_{1g} derived from Equation 5–2 can be used to determine β_s:

$$\beta_s = {}^1\!/_2\{-{}^1\!/_2[E(T_{2u}) + E(T_{1g})] + E(T_{2g})\} \qquad (5\text{–}6)$$

The energy parameter $E(T_{1u})_s$ is not included in this mean because of the uncertainty in the core-surface mixing parameter $\Delta E(T_{1u})$, obtained as outlined above, which must be subtracted from the value of $E(T_{1u})$, obtained from the actual computation.

An analogous but more complicated method can be used to analyze the computations of icosahedral boranes such as $B_{12}H_{12}^{2-}$. Thus, consider an icosahedrally weighted K_{12} complete graph, having 30 edges of unit weight corresponding to the icosahedron edges, 30 edges of weight m corresponding to the "meta" interaction of nonadjacent, nonantipodal vertex pairs, and 6 edges of weight p corresponding to the "para" interaction of the six icosahedron antipodal pairs (Figure 5–2). The spectrum of this graph is given in Table 5–3. Note that the T_{1u} core molecular orbital has a positive eigenvalue unless $p > \sqrt{5}(1 - m)$. Thus, with most likely values of the edge weights m and p, the icosahedrally weighted K_{12} graph has four positive eigenvalues, corresponding to the A_g and triply degenerate T_{1u} orbitals, rather than only the single positive eigenvalue characteristic of the unweighted K_{12} graph. Note that setting $m = 0$ and $p = 0$ gives the spectrum of the icosahedron (+5, $+\sqrt{5}$ three times, –1 five times, and $-\sqrt{5}$ three times as depicted in Figure 4–10) whereas setting $m = 1$ and $p = 1$ gives the spectrum of the K_{12} complete graph (+11, –1 eleven times).

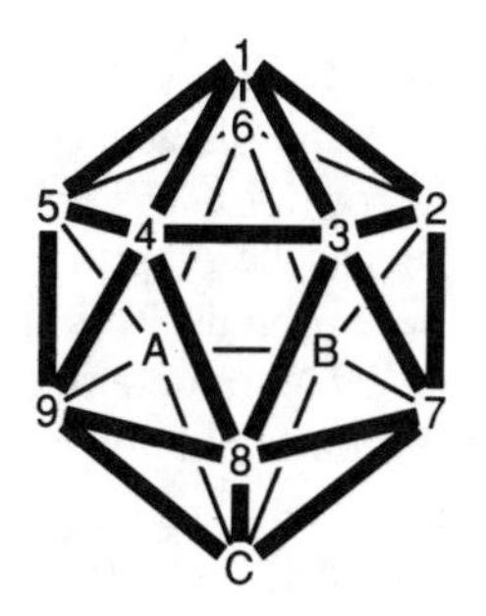

- "ortho" interactions (icosahedral edges): 12, 13, 14, 15, 16, 23, 26, 27, 2B, 34, 37, 38, 45, 48, 49, 56, 59, 5A, 6A, 6B, 78, 7B, 7C, 89, 8C, 9A, 9C, AB, AC, BC
- "meta" interactions: 17, 18, 19, 1A, 1B, 28, 29, 2A, 2C, 39, 3A, 3B, 3C, 47, 4A, 4B, 4C, 57, 58, 5B, 5C, 67, 68, 69, 6C, 79, 7A, 8A, 8B, 9B
- "para" interactions (antipodal pairs): 1C, 29, 3A, 4B, 57, 68

FIGURE 5–2. The "ortho", "meta", and "para" interactions in an icosahedron.

TABLE 5–3
Spectrum of the Icosahedrally Weighted K_{12} Graph

Eigenvalue	Degeneracy	Irreducible Representation	Mixes with Surface Orbital
$5 + 5m + p$	1	A_g	no
$\sqrt{5}(1-m)-p$	3	T_{1u}	yes
$-1-m-p$	5	H_g	yes
$-\sqrt{5}(1-m)-p$	3	T_{2u}	no

The spectrum of the icosahedrally weighted K_{12} graph indicates that, in the absence of core-surface and core-external orbital mixing, the energy parameters of the icosahedral core orbitals in $B_{12}H_{12}^{2-}$ relative to α are determined by the equations

$$E(A_{1g}) = (5 + 5m + p)\beta_c \qquad (5\text{–}7a)$$

$$E(T_{1u})_c = [\sqrt{5}(1-m) - p]\beta_c \qquad (5\text{–}7b)$$

$$E(\mathrm{H_g})_c = (-1 - m - p)\beta_c \tag{5–7c}$$

$$E(\mathrm{T_{2u}}) = [-\sqrt{5}(1 - m) - p]\beta_c \tag{5–7d}$$

in which β_c is the core orbital energy unit, m is the ratio of the *meta* (nonadjacent nonantipodal) to *ortho* (adjacent) interactions, and p is the ratio of *para* (antipodal) to *ortho* interactions. The only core icosahedral orbitals not subject to core-surface mixing are the A_g principal core orbital and the triply degenerate T_{2u} antibonding core orbitals. This leaves only the two Equations 5–7a and 5–7d for the energy parameters of the A_g and T_{2u} orbitals, respectively, to determine the three unknowns m, p, and β_c. The system is therefore underdetermined by one relationship so that an additional relationship between m, p, and/or β_c must be assumed before the necessary parameters can be extracted from the computed energy parameters. The arbitrary auxiliary assumption

$$m \approx 2p \tag{5–8}$$

is therefore introduced, allowing the following two equations for β_c and m to be derived from Equations 5–7a, 5–7d, and 5–8 and the decimal value 2.236 for $\sqrt{5}$:

$$\beta_c = \frac{E(\mathrm{A_{1g}})_c - (3.168)E(\mathrm{T_{2u}})_c}{12.083} \tag{5–9a}$$

$$m = 2p = \frac{2}{11}\left(\frac{E(\mathrm{A_{1g}})_c}{\beta_c} - 5\right) \tag{5–9b}$$

Analysis of the computations of Hoffmann and Lipscomb for $B_{12}H_{12}^{2-}$ using the 36 internal orbitals of the 12 boron atoms as the basis set suggests that the values of m and β_c obtained from a given set of computed molecular orbital energy parameters are relatively insensitive to the assumed relationship between m and p in the range $0 < p < m$. After determining m and β_c by Equations 5–9a and 5–9b, Equations 5–7b and 5–7c can be used to calculate hypothetical values for $E(\mathrm{T_{1u}})_c$ and $E(\mathrm{H_g})_c$ in the absence of core-surface mixing. Comparison of these values with the computed energy parameters for the T_{1u} and H_g *core* orbitals gives the parameters $\Delta E(\mathrm{T_{1u}})$ and $\Delta E(\mathrm{H_g})$ measuring the extent of core-surface mixing.

An approach similar to that used for octahedral $B_6H_6^{2-}$ can be used to estimate the surface orbital energy unit β_s for the icosahedral $B_{12}H_{12}^{2-}$, corresponding to a given set of computed molecular orbital energy parameters. The ideal surface orbital energy parameters for the icosahedron based on the symmetrical P_{12} group can be determined by the equations:

$$E(G_u) = E(H_g)_s = E(T_{1u})_s = \beta_s \quad (5\text{–}10a)$$

$$E(G_g) = E(H_u) = E(T_{1g}) = -\beta_s \quad (5\text{–}10b)$$

Reduction of the effective symmetry from the P_{12} symmetric group of the K_{12} complete graph with 12! operations to the actual I_h point group of the regular icosahedron with 120 operations destroys the equalities between the molecular orbital energy parameters in Equations 5–10a and 5–10b. On the basis of these equations and Equation 5–2, the following degeneracy-weighted mean of the energy parameters of the pure bonding surface orbital G_u of degeneracy 4 and the pure antibonding surface orbitals G_g, H_u, and T_{1g} of degeneracies 4, 5, and 3, respectively, can be used to determine β_s:

$$\beta_s = {}^1\!/_2\{-{}^1\!/_{12}[4E(G_g) + 5E(H_u) + 3E(T_{1g})] + E(G_u)\} \quad (5\text{–}11)$$

As in the case of the analogous calculation for octahedral $B_6H_6^{2-}$, the energy parameters $E(H_g)_s$ and $E(T_{1u})_s$ are not included in the mean, owing to uncertainties in estimating the core-surface mixing corrections $\Delta E(H_g)$ and $\Delta E(T_{1u})$.

Application of these methods for the topological analysis of various computations[1,4,5] on octahedral $B_6H_6^{2-}$ and icosahedral $B_{12}H_{12}^{2-}$ leads to the following observations:

1. The Hoffmann-Lipscomb LCAO-MO extended Hückel computations and the Armstrong-Perkins-Stewart self-consistent molecular orbital computations, both of which use Slater-type orbitals directly, give very similar values of β_c/β_s and t, particularly in the case of octahedral $B_6H_6^{2-}$.
2. The SCF-MO *ab initio* Gaussian 82 computations, which approximate Slater-type orbitals with a sum of Gaussians, give much lower values of both β_c/β_s and the non-adjacent core orbital interaction parameters (t for $B_6H_6^{2-}$ and m for $B_{12}H_{12}^{2-}$) than the computations using Slater orbitals directly. This indicates that the representation of Slater-type orbitals by a sum of Gaussians, as is typical in modern *ab initio* computations, leads to significantly weaker apparent core bonding, approximated more closely by deltahedral (D_n) rather than complete K_n topology, probably because Gaussian functions of the type $e^{\alpha r^2}$ fall off more rapidly at longer distances than Slater functions of the type $e^{-\zeta r}$.
3. The T_{1u} orbitals, which, if pure, would be non-bonding in octahedral (D_6) core bonding topology for $B_6H_6^{2-}$ and bonding in icosahedral (D_{12}) core bonding topology for $B_{12}H_{12}^{2-}$, become antibonding through core-surface mixing. Because of this feature, the simpler graph-theory derived model outlined in the previous chapter, using complete core bonding

topology where $G_c = K_n$ gives the correct numbers of bonding and antibonding orbitals for the deltahedral boranes, even though analyses of the type outlined in this chapter suggest that the complete graph K_n is a poor approximation of the actual G_c, corresponding to the computations using Gaussian orbitals.

Chapter 6

MOLECULAR AND IONIC TRANSITION METAL CARBONYL CLUSTERS

6.1 METAL CARBONYL VERTICES IN METAL CLUSTERS

Metal clusters[1] are compounds containing a metal skeleton in which three or more metal atoms are joined by metal-metal bonds. They may be characterized by the geometry of their metal skeletons such as the Fe_3 triangle in $Fe_3(CO)_{12}$, the Co_4 tetrahedron in $Co_4(CO)_{12}$, and the Rh_6 octahedron in $Rh_6(CO)_{16}$ (Figure 6–1).

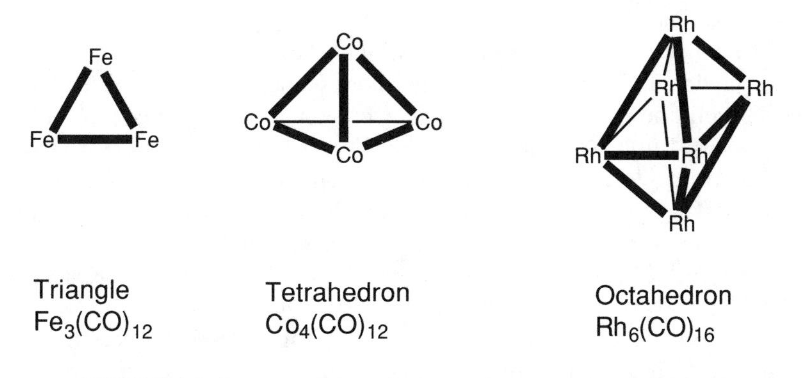

FIGURE 6–1. Some metal networks in simple metal carbonyl clusters.

Chapter 4 discusses polygonal hydrocarbons and polyhedral boranes in which the vertex atoms use only *s* and *p* orbitals leading to a four-orbital sp^3 manifold. Three of these four valence orbitals are used for the internal skeletal bonding, leaving only a single orbital for bonding to an external group such as hydrogen. The transition metal vertices in metal carbonyl clusters use *d* orbitals as well as *s* and *p* orbitals, leading to a nine-orbital sp^3d^5 manifold. In globally delocalized transition metal carbonyl clusters, three orbitals from each transition metal vertex are used for internal skeletal bonding, as is true in the polygonal hydrocarbons and the deltahedral boranes. This leaves $9 - 3 = 6$ valence orbitals on each transition metal vertex atom for bonding to one or generally more external ligands or groups, with the following being the most common in transition metal cluster vertices:

---

[1]B. F. G. Johnson, Ed., *Transition Metal Clusters*, Wiley-Interscience, Chichester, England, 1980.

1. A single external orbital bonding to a carbonyl group, an isocyanide ligand ($R–N^+≡C^-$:), or a trivalent phosphorus ligand (R_3P, $(R_2N)_3P$, $(RO)_3P$, etc.);
2. Three external orbitals forming a bond to all of the carbon atoms in a planar pentagonal (cyclopentadienyl), hexagonal (benzene), or heptagonal (tropylium or cycloheptatrienyl) ring;
3. A single external orbital containing a lone electron pair not bonded to an external group; such an external orbital is conveniently called a *non-bonding* external orbital. In all known cases, at least some of the external orbitals of heavy vertex atoms are non-bonding; most frequently the number of such non-bonding external orbitals is three.

The common types of vertex groups with transition metals (M) as vertex atoms are metal tricarbonyl vertices $M(CO)_3$ and cyclopentadienylmetal vertices C_5H_5M; both of these types of vertex groups have three non-bonding external orbitals.

The transition metals in metal carbonyls usually have the electronic configuration of the next rare gas, which corresponds to 18 valence electrons.[2] Thus, well-known mononuclear metal carbonyls having this favored 18 (valence) electron configuration include $Cr(CO)_6$, $Fe(CO)_5$, and $Ni(CO)_4$. A similar principle applies to transition metal vertices in metal carbonyl clusters. This provides a means for calculating the number of electrons provided by various transition metal vertex groups for the metal cluster skeleton; such electrons are called *skeletal electrons*. For example, consider an $Fe(CO)_3$ vertex group using three internal iron orbitals for the skeletal bonding, leaving six external iron orbitals requiring a total of 12 electrons to attain the favored 18-electron rare gas configuration for the iron atom. Of these required 12 electrons, two come from each of the three carbonyl groups, leaving six electrons to be provided by the iron atom. Since a neutral iron atom has eight valence electrons, this leaves 2 (= 8 – 6) electrons for the polygonal or polyhedral skeleton. Thus, an $Fe(CO)_3$ group is a donor of two skeletal electrons. Since a B–H vertex, as found in the deltahedral borane anions $B_nH_n^{2-}$ or the carboranes $C_2B_{n-2}H_n$ is also a donor of two skeletal electrons (see Chapter 4), an $Fe(CO)_3$ vertex is isoelectronic and isolobal to a B–H vertex. The isoelectronic analogy was first recognized by Wade[3] and subsequently extended by Hoffmann[4] to the concept of isolobality. Table 6–1 summarizes the skeletal electrons donated by important types of transition metal verticesand their isoelectronic and isolobal relationships to carbon and boron vertices. The hydrogen atoms in B–H and C–H vertices can be replaced

---

[2]J. E. Huheey, *Inorganic Chemistry: Principles of Structure and Reactivity*, 3rd Ed., Harper & Row, New York, 1983, 589.

[3]K. Wade, *Chem. Comm.*, 792, 1971.

[4]R. Hoffmann, *Angew. Chem. Int. Ed.,* **21**, 711, 1982.

by other monovalent groups, such as halogen, alkyl, aryl, nitro, cyano, etc.; the carbonyl groups in $M(CO)_3$ vertices can be replaced by other two-electron donor ligands, such as tertiary phosphines and isocyanides; and the hydrogen atoms on the C_5H_5 rings in the C_5H_5M vertices can be replaced partially or completely by other monovalent groups, notably methyl or trimethylsilyl. Also note that the details of the distribution of carbonyl groups on the metal cluster skeleton are rarely significant for electron counting, since both terminal and bridging carbonyls are two-electron donors, except for some rather rare types of dihapto bridging carbonyls which are almost never found in carbonyl derivatives of the middle to late transition metals forming clusters.

TABLE 6–1
Skeletal Electrons Donated by Transition Metal Vertices

Skeletal Electrons	B or C Vertex	$M(CO)_3$ Vertex	C_5H_5M Vertex
2	B–H	Fe, Ru, Os	Co, Rh, Ir
3	C–H	Co, Rh, Ir	Ni, Pd, Pt

Some metal carbonyl clusters have vertices of post-transition elements, such as phosphorus, sulfur, and their heavier congeners. Let us consider an alkylphosphinidene vertex R–P. The phosphorus atom of a *normal* alkylphosphinidene vertex uses only its *s* and *p* orbitals for chemical bonding, leading to four sp^3 tetrahedrally hybridized orbitals, one of which is an external orbital for bonding to the alkyl group and the remaining *three* of which are internal orbitals for the cluster skeletal bonding. However, in addition, *hypervalent* alkylphosphinidene vertices are also possible, with five $d(x^2–y^2)sp^3$ square pyramidally hybridized orbitals, one of which is an external orbital for bonding to the alkyl group and the remaining *four* of which are internal orbitals for the cluster skeletal bonding. Thus, the phosphorus atom in an alkylphosphinidene vertex can contribute either three or four internal orbitals to the cluster skeletal bonding, in contrast to boron or carbon vertices with energetically inaccessible *d* orbitals, which can contribute only three internal orbitals to the cluster skeletal bonding. Both normal and hypervalent alkylphosphinidene vertices contribute four skeletal electrons to the cluster bonding, since only one of the five phosphorus valence electrons is required for external bonding to the alkyl groups in either case. Thus, the counting of skeletal electrons in alkylphosphinidene metal clusters does *not* require previous determination whether the alkylphosphinidene vertices are normal or hypervalent. Also alkylarsinidene (RAs:) and bare sulfur (S:) vertices are isoelectronic and isolobal with alkylphosphinidene vertices. Bare phosphorus (P:) vertices are donors of three skeletal electrons, since two of the five phosphorus valence electrons are required for the external lone pair.

6.2 ELECTRON COUNTING IN ALTERNATIVE THEORIES FOR METAL CARBONYL CLUSTER SKELETAL BONDING

A number of theoretical approaches have been developed for describing metal carbonyl cluster skeletal bonding, leading to a number of different electron counting schemes. Some of these electron-counting schemes, including the graph-theory derived approach outlined in this book, count only the apparent skeletal electrons, whereas others count both skeletal and external electrons. Transition metal vertices are assumed to use three internal orbitals for skeletal bonding, leaving six external orbitals requiring 12 electrons for each vertex. Therefore, a globally delocalized deltahedron with n vertices which has $2n + 2$ skeletal electrons will have $12n + 2n + 2 = 14n + 2$ total electrons. For example, a six-vertex globally delocalized octahedron with $(2)(6) + 2 = 14$ skeletal electrons has $(14)(6) + 2 = 86$ total electrons. The polyhedral skeletal electron pair theory of Wade[5] and Mingos[6,7] uses essentially equivalent topological ideas but frequently uses total electron counts rather than skeletal electron counts, making the comparison between electron counts for polyhedra with different numbers of vertices but similar general topologies less obvious. However, the theories of Wade and Mingos arrive at the $2n + 2$ skeletal electron count for deltahedra with no vertices of degree 3, similar to the graph-theory derived approach used in this book.

Another method of interest is the topological electron-counting theory of Teo.[8,9] This theory starts from Euler's theorem for polyhedra in the form

$$e = v + f - 2 \tag{6–1}$$

where e, v, and f are the numbers of edges, vertices, and faces in the polyhedron in question (see Section 1.4 and Equation 1–9). Assuming that each atom on the surface of the polyhedron has the tendency to attain the 18-electron rare gas configuration as discussed above and that each edge can be considered as an edge-localized two-center two-electron metal-metal bond, the total electron count N for the cluster is

$$N = 18v - 2e \tag{6–2}$$

The N electrons will fill the $N/2$ energetically low-lying metal cluster valence molecular orbitals (CVMO) so that the number of CVMO's can be obtained by the equation

---

[5]K. Wade, *Adv. Inorg. Chem. Radiochem.*, **18**, 1, 1976.
[6]D. M. P. Mingos, *Nature (London), Phys. Sci.*, **236**, 99, 1972.
[7]D. M. P. Mingos, *Acc. Chem. Res.*, **17**, 311, 1984.
[8]B. K. Teo, *Inorg. Chem.*, **23**, 1251, 1984.
[9]B. K. Teo, G. Longoni, and F. R. K. Chung, *Inorg. Chem.*, **23**, 1257, 1984.

$$\text{CVMO} = N/2 = 9v - e = 8v - f + 2 \tag{6–3}$$

derived from Equations 6–1 and 6–2. However, for a delocalized metal cluster, such as globally delocalized metal deltahedra, not all metal-metal interactions can be considered to be edge localized. An adjustment factor X is introduced into Equation 6–3 reflecting delocalization leading to the equation

$$\text{CVMO} = N/2 = 9v - e + X = 8v - f + 2 + X \tag{6–4}$$

where X is the number of "extra" electron pairs "in excess" of the "18-electron rule." An alternative interpretation of X is that it is the number of "false" metal-metal bonds or, in molecular orbital terminology, the number of "missing" antibonding cluster orbitals, if each polyhedral edge is considered as an edge-localized bond.

Using this terminology, Teo has derived several rules for determining this adjustment factor X, of which the following are easiest to understand:

1. $X = 0$ for all polyhedra in which all vertices have degree 3, such as the tetrahedron, cube, and dodecahedron in Figure 4–1; these polyhedra are exactly those exhibiting edge-localized bonding in accord with the "matching rule" in Section 4.2;
2. Capping a face of a polyhedron having n edges leads to an increase in X by $n - 3$. Hence, capping a triangular, quadrilateral (including square), and pentagonal faces increases X by 0, 1, and 2, respectively;
3. $X = 0$ for all pyramids;
4. The values of X for bipyramids can be determined by capping the corresponding pyramids or by considering the molecular orbitals of the polygon forming the "belt" of the bipyramid, leading to $X = 0$ for the trigonal bipyramid, $X = 1$ for a tetragonal bipyramid, and $X = 2$ for a pentagonal bipyramid. Elongation of a trigonal bipyramid can increase its X value from 0 to 2.

These rules have been justified by considering the molecular orbitals of the polyhedra in question.[8,9]

6.3 ELECTRON-POOR METAL CARBONYL CLUSTERS

Chapter 4 discusses the skeletal electron counts for the polyhedral boranes. Deltahedral boranes with n vertices, in which all vertices have degrees 4, 5, or (rarely) 6, have the favored $2n + 2$ skeletal electrons. Electron-rich polyhedral boranes having more than $2n + 2$ skeletal electrons are constructed from polyhedra having at least one non-triangular face. Such polyhedra are also found in metal carbonyl clusters, although metal carbonyl clusters based on delocalized electron-rich polyhedra similar to the electron-rich polyhedral boranes (e.g., *nido* boranes with $2n + 4$ skeletal electrons or *arachno* boranes with $2n + 6$ skeletal electrons) are rather rare. Although transition metal vertices in metal carbonyl clusters may be isoelectronic and isolobal with boron and carbon vertices (see Table 6–1), the presence of *d* orbitals in addition to *s* and *p* orbitals on the transition metal vertices leads to some additional possibilities. Thus, a transition metal vertex can use more than three of its valence orbitals for (internal) skeletal bonding, so that orbitals can be available for edge-localized bonding from vertices of degrees 4 or 5 (see the "matching rule" in Section 4.2). In addition, the availability of *d* orbitals on transition metal vertices also allows for electron-poor transition metal clusters with *n* vertices having *less* than $2n + 2$ *apparent* skeletal electrons. Such electron-poor clusters form deltahedra containing tetrahedral chambers, i.e., deltahedra with one or more vertices of degree 3. The simplest examples of such deltahedra are the capped tetrahedra, of which the trigonal bipyramid (i.e., the monocapped tetrahedron) with five vertices is the smallest (Figure 6–2). The capped tetrahedra consist of a series of fused tetrahedral chambers with faces in common. An example of a cluster based on a bicapped tetrahedron is $Os_6(CO)_{18}$, which has 12 (i.e., $2n$) skeletal electrons. The simplest deltahedron in which tetrahedral chambers do not occupy the whole volume of the polyhedron is the capped octahedron with 7 vertices; this polyhedron is found in $Rh_7(CO)_{16}^{3-}$, which has 14 (i.e. $2n$) skeletal electrons (Figure 6-2).

The following properties of tetrahedral chambers in capped deltahedra are significant:

1. Tetrahedral chambers in capped deltahedra require localized bonding like isolated tetrahedra. Tetrahedral chambers formed by capping deltahedra without degree 3 vertices may be regarded as islands of localization in an otherwise delocalized system;
2. Atoms at vertices of triangular faces being capped require more than three internal orbitals oriented inwards towards the cluster polyhedron. Therefore, such atoms cannot be light atoms (e.g., boron or carbon) having only *s* and *p* orbitals in an sp^3 manifold, since this total of four valence orbitals cannot all be oriented on the same side of the plane bisecting the light atom (i.e., in the same "halfspace"). For this reason the analogies between the transition metal vertices in metal clusters and the boron vertices in polyhedral boranes outlined in Table 6–1 break down;
3. The capping vertex contributes the usual number of skeletal electrons but no skeletal orbitals. Therefore, capping is a good remedy for "electron poverty."

There are two opposite or dual processes for converting closed deltahedra with *n* vertices, which require $2n + 2$ skeletal electrons into polyhedra

FIGURE 6–2. Examples of deltahedra with tetrahedral chambers (i.e., deltahedra with degree 3 vertices) found in electron poor metal carbonyl clusters.

appropriate for systems with a larger or smaller number of skeletal electrons relative to the number of vertices. For electron-rich polyhedra having more than $2n + 2$ skeletal electrons, the appropriate process is either polyhedral excision or the equivalent polyhedral puncture. In polyhedral excision, a vertex and all of its incident edges are removed so that more electrons than bonding orbitals are lost. For electron-poor polyhedra having less than $2n + 2$ skeletal electrons, the appropriate process is polyhedral capping, in which a triangular face is capped with a new vertex to add electrons to the structure without adding bonding orbitals.

The bicapped tetrahedron (Figure 6–2) consisting of three fused tetrahedral chambers (i.e., an "analogue" of anthracene based on three-dimensional tetrahedra rather than two-dimensional hexagons), such as that found in $Os_6(CO)_{18}$, is a good example for illustrating the concept of apparent skeletal electron counting for electron-poor polyhedra. Since the bicapped tetrahedron has 12 edges, 24 skeletal electrons are required for edge-localized bonding as suggested by the three tetrahedral chambers. However, in a model assuming such edge-localized bonding, there are six "extra" internal orbitals arising from the two vertices of degree 4 and the two vertices of degree 5 (i.e., $(2)(4–3) + (2)(5–3) = 6$). These extra internal orbitals provide 12 of these 24 skeletal electrons. Therefore, the *apparent* skeletal electron count for the bicapped tetrahedron is $24 - 12 = 12$. Five of the nine valence orbitals in the sp^3d^5 manifold of the degree 5 osmium vertices can be directed inward towards the edges of the bicapped tetrahedron to participate in skeletal rather than external bonding, leading to the matching of the numbers of internal orbitals and vertex degrees required for edge localized bonding.

6.4 ELECTRON COUNTING IN METAL CARBONYL CLUSTERS

This section provides examples of skeletal electron counting in metal clusters based on various polyhedra having from four to twelve vertices.

A. TETRAHEDRA (FIGURE 6–1):

The normal $M(CO)_3$ vertices in the tetrahedral metal clusters $M_4(CO)_{12}$ (M = Co, Rh, Ir) provide three skeletal electrons each, leading to the (4)(3) = 12 skeletal electrons required for two-electron two-center bonds along each of the six edges of the M_4 tetrahedron.

B. TRIGONAL BIPYRAMID (FIGURE 6–2):

The cluster $Os_5(CO)_{16}$ has a trigonal bipyramidal structure[10] indicative of some edge-localization associated with the degree 3 apical vertices. It has 18 skeletal electrons according to the following skeletal electron counting scheme:

2 degree 3 axial $Os(CO)_3$ vertices:	
$2[2(3+3) - 10] = (2)(2) =$	4 electrons
3 degree 4 equatorial $Os(CO)_3$ vertices:	
$3[2(3+4) - 10] = (3)(4) =$	12 electrons
Extra CO group	2 electrons
Total skeletal electrons	*18 electrons*

These 18 skeletal electrons correspond to edge-localized bonding with a two-electron bond along each of the nine edges of the trigonal bipyramid. Subtraction of a total of six skeletal electrons for the fourth internal orbitals of each of the three equatorial osmium atoms in $Os_5(CO)_{16}$ from these actual skeletal electrons gives the 12 apparent skeletal electrons which would be obtained for a trigonal bipyramid if each of the five vertices are assumed to be normal vertices contributing three internal orbitals to the skeletal bonding. Thus, erroneously considering $Os_5(CO)_{16}$ as a globally delocalized system after regarding $Os(CO)_3$ vertices as using three internal orbitals regardless of their position in the trigonal bipyramid leads to the following scheme for counting the *apparent* skeletal electrons:

5 degree 3 $Os(CO)_3$ vertices: (5)(2) =	10 electrons
Extra CO group	2 electrons
Apparent skeletal electrons:	*12 electrons*

These 12 skeletal electrons would be considered to correspond to the $2n + 2$ skeletal electrons required for a five-vertex globally delocalized structure (i.e., $n = 5$). However, this correspondence is only fortuitous, since consideration of a trigonal bipyramid as a delocalized rather than an edge-localized cluster contradicts principles that are necessary to explain the electron counts in other cluster structures

---

[10]C. R. Eady, B. F. G. Johnson, J. Lewis, B. E. Reichert, and G. M. Sheldrick, *Chem. Comm.*, 271, 1976.

Another trigonal bipyramidal metal carbonyl cluster is $Rh_5(CO)_{16}^-$ (ref. 11) which has 16 apparent skeletal electrons using the following electron counting scheme:

5 $Rh(CO)_3$ vertices: (5)(3) =	15 electrons
–1 charge on ion	1 electron
Total apparent skeletal electrons:	*16 electrons*

The 16 apparent skeletal electrons for $Rh_5(CO)_{15}^-$ as compared with the 12 apparent skeletal electrons for $Os_5(CO)_{16}$ obtained as noted above indicates that the skeleton of the rhodium cluster has four more bonding electrons than that of the osmium cluster. This anomaly can be rationalized if the two axial vertices in the $Rh_5(CO)_{15}^-$ cluster are regarded as anomalous by contributing only two internal orbitals to the trigonal bipyramidal skeletal bonding rather than the normal three internal orbitals. This has the following consequences:

1. In a trigonal bipyramid, the degrees of the two axial vertices and those of the three equatorial vertices are 3 and 4, respectively. Therefore, if the axial vertex atoms contribute 2 internal orbitals and the equatorial vertex atoms contribute 3 internal orbitals, there is a mismatch between the degrees of each of the five vertices of the trigonal bipyramid and the numbers of internal orbitals contributed by the vertex atoms. This suggests the possibility of delocalized bonding in the $Rh_5(CO)_{15}^-$ trigonal bipyramid.
2. An $Rh(CO)_3$ vertex using only two internal orbitals then has seven external orbitals, three of which are used for bonding to CO groups. This leaves four external orbitals for eight of the nine electrons furnished by a neutral rhodium atom, thereby leaving only one electron for the skeletal bonding. Therefore, an anomalous $Rh(CO)_3$ vertex, contributing only two internal orbitals to the cluster polyhedron, is a donor of only one skeletal electron, in contrast to a normal $Rh(CO)_3$ vertex, contributing the normal three internal orbitals, which is a donor of three skeletal electrons. This leads to the following skeletal electron counting scheme for $Rh_5(CO)_{15}^-$, in which the apical $Rh(CO)_3$ vertices use only two internal orbitals:

2 apical $Rh(CO)_3$ vertices using 2 internal orbitals:	
(2)(1) =	2 electrons
3 equatorial $Rh(CO)_3$ vertices using 3 internal orbitals:	
(3)(3) =	9 electrons
–1 charge on ion	1 electron
Total skeletal electrons:	*12 electrons*

---

[11] A. Fumagalli, T. F. Koetzle, F. Takusagawa, P. Chini, S. Martinengo, and B. T. Heaton, *J. Am. Chem. Soc.*, **102**, 1740, 1980.

The resulting 12 skeletal electrons correspond to the $2n + 2$ skeletal electrons required for a globally delocalized structure when $n = 5$.

3. The two internal orbitals from each of the axial rhodium vertices in $Rh_5(CO)_{15}^-$ are best regarded as twin internal orbitals. The axial rhodium vertices therefore do *not* have a unique internal orbital available for the multicenter core bonding, which therefore uses only the three unique internal orbitals of the three equatorial rhodium atoms. However, the Rh_5 cluster in $Rh_5(CO)_{15}^-$ has the required six skeletal bonding orbitals for the observed 12 skeletal electrons, since five skeletal bonding orbitals arise from the pairwise interaction of the 10 twin internal orbitals of the five rhodium atoms, and the sixth skeletal bonding orbital arises from the three-center overlap of the three unique internal orbitals of the three equatorial rhodium atoms (i.e., overlap according to the topology of the K_3 complete graph).
4. Since the axial rhodium atoms in $Rh_5(CO)_{15}^-$ do not participate in the multicenter core bonding, the axial-axial Rh–Rh distance in $Rh_5(CO)_{15}^-$ should be long relative to the axial-axial distance in trigonal bipyramids in which the metal atoms contribute three internal orbitals. In other words, trigonal bipyramids in which the axial atoms contribute only two internal orbitals should be more elongated than trigonal bipyramids in which the axial atoms contribute the usual 3 internal orbitals. An elongation factor E for *homonuclear* trigonal bipyramids can be estimated from bond distances determined by X-ray diffraction using the equation $E = \hat{r}_{ae}/\hat{r}_{ee}$ in which $\hat{r}_{ae}$ is the average of the six axial-equatorial metal-metal bond distances and $\hat{r}_{ee}$ is the average of the three equatorial-equatorial metal-metal bond distances. The parameter E has the values 1.08, 1.10, and 1.19 for the isoelectronic series of homonuclear trigonal bipyramids $Rh_5(CO)_{15}^-$, $Rh_5(CO)_{14}I^{2-}$ (ref. 12), and $Ni_5(CO)_{12}^{2-}$ (ref. 13), respectively, which has 12 skeletal electrons if the two axial vertices contribute two internal orbitals each and the three equatorial vertices contribute three internal orbitals each. However, the parameter E has the value of only 0.99 for $Os_5(CO)_{16}$, which has 12 skeletal electrons if all five vertices are normal vertices contributing three internal orbitals each.

C. OCTAHEDRON (FIGURE 6–1):

The simple octahedral metal carbonyl clusters $M_6(CO)_{16}$ (M = Co, Rh, Ir) have the 14 skeletal electrons (= $2n + 2$ for $n = 6$) required for a globally delocalized octahedron as determined by the following electron counting scheme:

---

[12] S. Martinengo, G. Ciani, and A. Sironi, *Chem. Comm.*, 1059, 1979.

[13] G. Longoni, P. Chini, L. D. Lower, and L. F. Dahl, *J. Am. Chem. Soc.*, **97**, 5034, 1975.

6 normal $M(CO)_3$ vertices: (6)(3) =	18 electrons
Deficiency of 2 CO groups: (–2)(2) =	–4 electrons
Total skeletal electrons:	*14 electrons*

A closely related octahedral metal cluster anion is $Ni_6(CO)_{12}^{2-}$ (ref. 14) which also has the 14 skeletal electrons required for a globally delocalized octahedron by the following electron-counting scheme:

6 normal $Ni(CO)_2$ vertices: (6)(2) =	12 electrons
–2 charge of anion:	2 electrons
Total skeletal electrons:	*14 electrons*

A number of alkylphosphinidenemetal carbonyl clusters containing *trans*-P_2M_4 octahedra are known which may contain either 14 or 16 apparent skeletal electrons.[15] The clusters of the type $(RP)_2Fe_4(CO)_{11}$ contain the 14 apparent skeletal electrons required for a globally delocalized octahedron using the following electron counting scheme:

2 RP vertices: (2)(4) =	8 electrons
4 $Fe(CO)_3$ vertices: (4)(2) =	8 electrons
"Deficiency" of one CO group:	–2 electrons
Total apparent skeletal electrons:	*14 electrons*

The clusters $(RP)_2Fe_4(CO)_{12}$ with an extra carbonyl group then contain 16 apparent skeletal electrons rather than the 14 apparent skeletal electrons for a globally delocalized octahedron. These 16 *apparent* skeletal electron P_2M_4 octahedral clusters are formulated as 24 *actual* skeletal electron *edge-localized* octahedral clusters having hypervalent alkylphosphinidene vertices. The eight extra skeletal electrons arise from the single extra internal orbitals on each of the four iron vertices required for these iron vertices each to have the four internal orbitals required for edge-localized bonds to the adjacent four atoms (two iron atoms and two phosphorus atoms) of the P_2Fe_4 octahedron, rather than only the three internal orbitals required for globally delocalized bonding. The existence of the edge-localized $(RP)_2Fe_4(CO)_{12}$ clusters is an excellent example of the ability for hypervalent alkylphosphinidene vertices to stabilize edge-localized bonding at degree 4 vertices. Thus the 16 apparent skeletal electron P_2Fe_4 clusters have been shown by X-ray diffraction to have the octahedral geometries for such edge-localized structures rather than the pentagonal pyramidal geometries expected for *nido* delocalized electron-rich $2n + 4 = 16$ skeletal electron systems for $n = 6$ (Section 4.4).

---

[14]J. C. Calabrese, L. F. Dahl, A. Cavalieri, P. Chini, G. Longoni, and S. Martinengo, *J. Am. Chem. Soc.*, 96 , 2616, 1974.

[15]R. B. King, *New J. Chem.*, **13**, 293, 1989.

This analysis indicates that edge-localized octahedral alkylphosphinidene metal carbonyl clusters have two more apparent skeletal electrons than the corresponding globally delocalized clusters. The black globally delocalized $(RP)_2Fe_4(CO)_{11}$ reversibly add CO to form the red edge-localized $(RP)_2Fe_4(CO)_{12}$ clusters[16] analogous to the hydrogenation of the globally delocalized benzene to give cyclohexane. In addition, the NMR proton chemical shift of the P-methyl groups in the globally delocalized $(CH_3P)_2Fe_4(CO)_{11}$ is δ 0.98 in contrast to that of δ 3.34 for the edge-localized $(CH_3P)_2Fe_4(CO)_{12}$, suggesting a "surface current" effect similar to the ring current effects in the NMR spectra of benzene and other planar two-dimensional globally delocalized structures.[17]

D. BICAPPED TETRAHEDRON (FIGURE 6–2):

The bicapped tetrahedron, which may alternatively be regarded as three fused tetrahedra, has 6 vertices, 12 edges, and 8 faces, like the regular octahedron. However, the bicapped tetrahedron has two vertices each of degrees 3, 4, and 5, whereas each of the six vertices of the octahedron has degree 4. The bicapped tetrahedron is only rarely found in metal carbonyl clusters. However, the discovery of the bicapped tetrahedron cluster $Os_6(CO)_{18}$ in the early days of metal cluster chemistry (1973)[18] was a key result for the development of metal cluster skeletal bonding theories, since previously all six-vertex metal clusters were expected to be octahedra. The $Os_6(CO)_{18}$ cluster may be formulated as a 24 skeletal electron system using the following electron counting scheme:

2 degree 3 (capping) $Os(CO)_3$ vertices: (2)(2) =	4 electrons
2 degree 4 $Os(CO)_3$ vertices: 2[2(3+4) – 10] = (2)(4) =	8 electrons
2 degree 5 $Os(CO)_3$ vertices: 2[2(3+5) – 10] = (2)(6) =	12 electrons
Total skeletal electrons:	*24 electrons*

Note that for each additional internal orbital provided by an $Os(CO)_3$ vertex, the number of skeletal electrons provided increases by two, corresponding to the incorporation of a previously non-bonding external electron pair into the skeletal bonding through the additional internal orbital. These 24 skeletal electrons from the above electron counting scheme correspond to edge-localized bonding with a two-electron bond along each of the 12 edges of the bicapped tetrahedron. Subtracting a total of 12 skeletal electrons for the "extra" internal orbitals of the osmium atoms at the two degree 5 vertices and the two degree 4 vertices from these 24 actual skeletal electrons gives the 12

---

[16]T. Jaeger, S. Aime, and H. Vahrenkamp, *Organometallics*, **5**, 245, 1986.

[17]I. Ando and G. A. Webb, *Theory of NMR Parameters*, Academic Press, New York, 1983, 69.

[18]R. Mason, K. M. Thomas, and D. M. P. Mingos, *J. Am. Chem. Soc.*, **95**, 3802, 1973.

apparent skeletal electrons obtained by erroneously treating $Os_6(CO)_{18}$ as a globally delocalized system with each of the osmium vertices providing the normal three internal orbitals. The bicapped tetrahedral rather than octahedral geometry for $Os_6(CO)_{18}$ can be rationalized on the basis that the electron-poor $Os_6(CO)_{18}$ has only 12 ($= 2n$ for $n = 6$) rather than the 14 skeletal electrons ($= 2n + 2$ for $n = 6$) required for an octahedron with globally delocalized bonding. This simplified electron counting procedure is useful as a crude device for identifying electron-poor systems having less than $2n + 2$ apparent skeletal electrons. However, the above more detailed electron counting for $Os_6(CO)_{18}$ can relate its skeletal electron count more precisely to edge-localized bonding in a specific polyhedron having tetrahedral chambers.

E. CAPPED OCTAHEDRA (FIGURE 6–3):

Metal carbonyl clusters, mainly of osmium, are known, in which one or more of the faces of a metal octahedron are capped with additional metal atoms to form metal clusters having 7, 8, 9, and 10 metal vertices of which 1, 2, 3, or 4 vertices, respectively, are capping vertices (Figure 6–3). Thus, the osmium carbonyl clusters depicted in Figure 6–3 may all be regarded as derivatives of a homologous series of clusters of the type $Os_{6+p}(CO)_{19+2p}$ formed by capping a central Os_6 octahedron with p μ_3-$Os(CO)_3$ caps where $0 \leq p \leq 4$. Multiple capping occurs so that pairs of faces of the original octahedron sharing an edge are never both capped, since otherwise some vertices of the central Os_6 octahedron would need six internal orbitals: three for the globally delocalized bonding in the central Os_6 octahedron and three for the localized bonds to the μ_3-$Os(CO)_3$ caps. Steric considerations concerning the orientations of the nine orbitals of the sp^3d^5 manifold of an osmium vertex suggest that five might be the maximum possible number of internal orbitals which can be contributed to the surface atoms of an osmium polyhedron. This suggests that the Os_{10} polyhedron depicted in Figure 6–3 is likely to be the largest Os_{6+p} polycapped octahedron which can be formed by capping a total of p faces of a central Os_6 octahedron with μ_3-$Os(CO)_3$ groups.

First, consider the monocapped octahedral cluster $Os_7(CO)_{21}$ (ref. 19). The seven $Os(CO)_3$ vertices, if considered artificially to be normal vertices contributing three internal orbitals each, give $Os_7(CO)_{21}$ a total of 14 apparent skeletal electrons corresponding to a $2n$ electron-poor system for $n = 7$, but corresponding to the required skeletal electrons for the globally delocalized octahedral cavity comprising six of the seven $Os(CO)_3$ vertices.

In considering $Os_7(CO)_{21}$ in more detail and constructing its Os_7 capped octahedron by fusion of an octahedron with a tetrahedron by sharing a(triangular) face, the seven $Os(CO)_3$ vertices can be divided into the following three categories:

---

[19] C. R. Eady, B. F. G. Johnson, J. Lewis, R. Mason, P. B. Hitchcock, and K. M. Thomas, *Chem. Comm.*, 385, 1977.

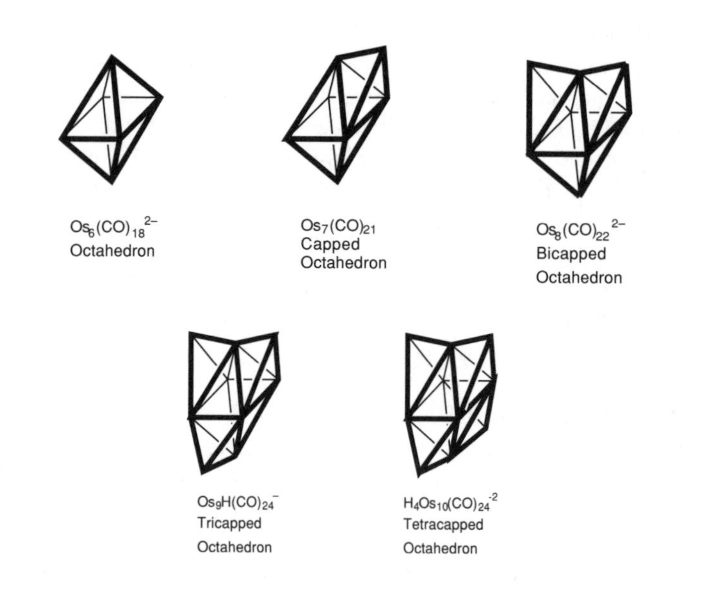

FIGURE 6–3. Capping octahedra with examples from osmium carbonyl cluster chemistry.

1. The single $Os(CO)_3$ vertex of degree 3 capping the octahedron and therefore a vertex only of the tetrahedron;
2. The three $Os(CO)_3$ vertices of degree 5 in the face of the octahedron being capped, and therefore vertices of both the tetrahedron and octahedron;
3. The remaining three $Os(CO)_3$ vertices of degree 4, which are vertices only of the octahedron.

The $Os(CO)_3$ vertex of Type 1. clearly provides three internal orbitals, the $Os(CO)_3$ vertices of Type 2. provide three internal orbitals for the globally delocalized octahedron and a fourth internal orbital for an edge-localized bond to the $Os(CO)_3$ cap (Type 1.), and the $Os(CO)_3$ vertices of Type 3 provide only the three internal orbitals required for a globally delocalized octahedron. The four $Os(CO)_3$ vertices of Types 1. and 3., providing only three internal orbitals, are donors of two skeletal electrons each, whereas the three $Os(CO)_3$ vertices of Type 2., providing a total of four internal orbitals, are donors of four skeletal electrons each, leading to $(4)(2) + (3)(4) = 20$ skeletal electrons. Of these 20 skeletal electrons, 14 are used for the globally delocalized octahedron, and the remaining six skeletal electrons are used for the three two-electron two-center bonds from the capping $Os(CO)_3$ vertex (Type 1. above) to the three $Os(CO)_3$ vertices (Type 2. above) of the face being capped. Thus, a capped octahedral metal carbonyl cluster is partially delocalized and partially localized: the bonding in the octahedral cavity is delocalized, like that in an isolated globally delocalized octahedron (e.g., $M_6(CO)_{16}$ (M = Co, Rh, Ir) discussed above), whereas the bonding in the tetrahedral cavity is edge-localized like that in an isolated edge-localized tetrahedron (e.g., $M_4(CO)_{12}$ (M = Co, Rh, Ir) discussed above).

Another example of a capped octahedral metal carbonyl cluster is the rhodium carbonyl anion $Rh_7(CO)_{16}^{3-}$ (ref. 20), which is isoelectronic with $Os_7(CO)_{21}$.

Continuation of the capping of a metal cluster octahedron leads ultimately to a cluster containing 10 metal atoms at the vertices of a tetracapped octahedron, in which the four faces of the central octahedron are so chosen not to share edges. Thus all 12 edges of the central octahedron are all edges of tetrahedra in the M_{10} cluster. If the osmium carbonyl clusters depicted in Figure 6–3 are each treated as globally delocalized clusters with all vertices considered as contributing three internal orbitals to the skeletal bonding, each of the clusters has the 14 apparent skeletal electrons required for a globally delocalized central octahedron using the following electron-counting schemes:

1. Bicapped octahedron, $Os_8(CO)_{22}^{2-}$ (ref. 21):

8 $Os(CO)_3$ vertices: (8)(2) =	16 electrons
"Deficiency" of 2 CO groups: (–2)(2) =	–4 electrons
–2 charge on ion:	2 electrons
Total apparent skeletal electrons:	*14 electrons*

2. Tricapped octahedron, $Os_9H(CO)_{24}^{-}$(ref. 22):

9 $Os(CO)_3$ vertices: (9)(2) =	18 electrons
"Deficiency" of 3 CO groups: (–3)(2) =	–6 electrons
H atom:	1 electron
–1 charge on ion:	1 electron
Total apparent skeletal electrons:	*14 electrons*

3. Tetracapped octahedron, $Os_{10}H_4(CO)_{24}^{-2}$ (ref. 23):

10 $Os(CO)_3$ vertices: (10)(2) =	20 electrons
"Deficiency" of 6 CO groups: (–6)(2) =	–12 electrons
4 H atoms:	4 electrons
–2 charge on ion:	2 electrons
Total apparent skeletal electrons:	*14 electrons*

---

[20] V. G. Albano, P. L. Bellon, and G. F. Ciani, *Chem. Comm.*, 1024, 1969.

[21] P. F. Jackson, B. F. G. Johnson, J. Lewis, and P. R. Raithby, *Chem. Comm.*, 60, 1980.

[22] A. J. Amoroso, B. F. G. Johnson, J. Lewis, P. Raithby, and W. T. Wong, *Chem. Comm.*, 814, 1991.

[23] D. Braga, J. Lewis, B. F. G. Johnson, M. McPartlin, W. J. H. Nelson, and M. D. Vargas, *Chem. Comm.*, 241, 1983.

F. TRICAPPED TRIGONAL PRISM (FIGURE 4–5):

The tricapped trigonal prism is the nine-vertex deltahedron having only degree 4 and degree 5 vertices and thus is the nine-vertex deltahedron without any tetrahedral chambers. As such, it is the deltahedron found in the globally delocalized nine-vertex deltahedral borane anion $B_9H_9^{2-}$.[24] An example of a metal carbonyl cluster basedon the tricapped trigonal prism is the alkylphosphinidene rhenium carbonyl $(\mu_4\text{-}CH_3P)_3Re_6(CO)_{18}$, which has hypervalent alkylphosphinidine groups at the three vertices of degree 4 and $Re(CO)_3$ vertices at the six vertices of degree 5 of the tricapped trigonal prism. If globally delocalized bonding with each vertex contributing three internal orbitals is assumed for $(\mu_4\text{-}CH_3P)_3Re_6(CO)_{18}$, as is the case for $B_9H_9^{2-}$, then the apparent skeletal electron count is obtained as follows:

3 hypervalent CH_3P vertices: (3)(4) =	12 electrons
6 $Re(CO)_3$ vertices: (6)[7 – (6–3)(2)] = (6)(1) =	6 electrons
Total apparent skeletal electrons:	*18 electrons*

These 18 skeletal electrons are two short of the 20 ($= 2n + 2$ for $n = 9$) skeletal electrons required for a globally delocalized trigonal prism. However, the actual skeletal electron count of $(\mu_4\text{-}CH_3P)_3Re_6(CO)_{18}$ is consistent with edge-localized bonding as follows:

3 hypervalent degree 4 CH_3P vertices: (3)(4) =	12 electrons
6 $Re(CO)_3$ vertices using 5 internal orbitals:	
(6)[7–(4–3)(2)] = (6)(5)=	30 electrons
Total actual skeletal electrons:	*42 electrons*

These 42 actual skeletal electrons are exactly the number required for edge-localized bonding along the 21 edges of the tricapped trigonal prism.

G. ICOSAHEDRON (FIGURE 4–5):

Icosahedral metal carbonyl clusters are known in which some of the vertices are alkylarsinidene vertices. Two examples are the $Ni_{10}As_2$ icosahedron $(\mu_5\text{-}CH_3As)_2Ni_{10}(CO)_{18}^{2-}$ and the Ni_9As_3 icosahedron $(\mu_5\text{-}CH_3As)_3Ni_9(CO)_{15}^{2-}$, each with the required $26 = 2n + 2$ skeletal electrons for globally delocalized polyhedra with 12 vertices ($n = 12$).[25] Both of these icosahedral nickel-arsenic clusters are isoelectronic with the $B_{12}H_{12}^{2-}$ anion. Note that since *all* vertices of a regular icosahedron have degree 5, matching the vertex degrees with either a normal alkylarsinidene vertex having three internal orbitals or a hypervalent alkylarsinidene vertex having four

---

[24]F. Klanberg and E. L. Muetterties, *Inorg. Chem.*, **5**, 1955, 1966.

[25]D. R. Rieck, R. A. Montag, T. S. McKechnie, and L. F. Dahl, *J. Am. Chem. Soc.*, **108**, 1330, 1986.

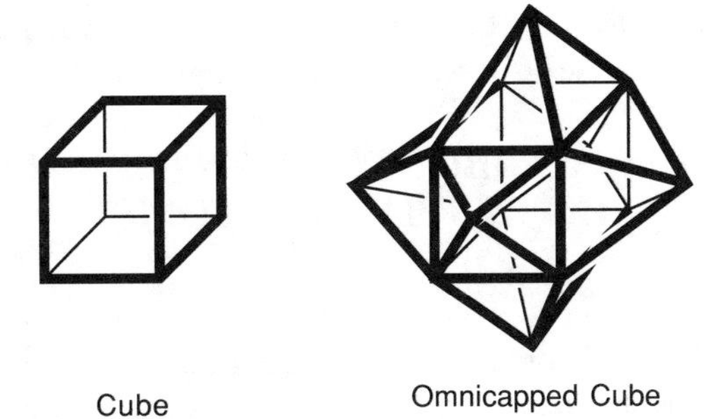

FIGURE 6–4. The cube and omnicapped cube. In $(\mu_4\text{-}C_6H_5P)_6Ni_8(CO)_8$ the nickel atoms are located at the eight degree 6 vertices of the omnicapped cube, and the phosphorus atoms are located at the six degree 4 "caps" of the omnicapped cube.

internal orbitals is not possible, so that the skeletal bonding in icosahedral clusters containing alkylphosphinidene or alkylarsinidene vertices *must* be globally delocalized. Other icosahedral metal carbonyl clusters contain interstitial atoms in the center of the icosahedron and therefore are treated later (Section 6.4).

H. CUBE/OMNICAPPED CUBE (FIGURE 6–4):

An interesting 14-atom P_6Ni_8 cluster is $(\mu_4\text{-}C_6H_5P)_6Ni_8(CO)_8$, in which all six faces of a central Ni_8 cube are capped by phenylphosphinidene vertices.[26] This can be treated as a cubic Ni_8 cluster, with C_6H_5P caps on each of the six faces of the cube leading to the following electron counting scheme:

8 normal NiCO vertices: (8)[10 – (6–1)(2)] = (8)(0) =	0 electrons
6 face-bridging C_6H_5 groups: (6)(4) =	24 electrons
Total apparent skeletal electrons:	*24 electrons*

The resulting 24 skeletal electrons correspond to two-electron two-center bonds along each of the 12 edges of the Ni_8 cube. An alternative, but equivalent, treatment of $(\mu_4\text{-}C_6H_5P)_6Ni_8(CO)_8$ considers all 14 vertex atoms P_6Ni_8 as forming an "omnicapped cube," which is a 14-vertex deltahedron which necessarily has 36 edges and 24 faces (Figure 6–4). The actual skeletal electrons are then counted as follows:

6 hypervalent (degree 4) C_6H_5 vertices: (6)(4) =	24 electrons
8 NiCO vertices using 6 internal orbitals:	
(8)[10 – (9–6–1)(2)] = (8)(6) =	48 electrons
Total skeletal electrons:	*72 electrons*

[26]L. D. Lower and L. F. Dahl, *J. Am. Chem. Soc.*, **98**, 5046, 1976.

These 72 actual skeletal electrons are exactly the number required for edge-localized bonds along the 36 edges of the omnicapped cube.

6.5 METAL CARBONYL CLUSTERS HAVING INTERSTITIAL ATOMS

Many metal carbonyl clusters have interstitial atoms or groups located in the center of the polyhedron. Such interstitial atoms may be a light atom such as carbon, a post-transition element such as antimony, or a transition metal. Interstitial atoms usually provide all of their valence electrons as skeletal electrons, since all of their valence orbitals are necessarily internal orbitals because of the location of the interstitial atom in the center of the polyhedron. Exceptions to this rule may occur (although rather rarely) when some of the valence electrons of the interstitial atom are in orbitals of symmetries which cannot mix with any of the molecular orbitals arising from the polyhedral skeletal bonding.

Interstitial atoms create certain volume requirements for the surrounding polyhedron.[27] Thus, an interstitial carbon atom cannot fit into a tetrahedron but fits into an octahedron as exemplified by $Ru_6(CO)_{17}C$. An interstitial transition metal cannot fit into an octahedron but fits into a larger polyhedron, such as a twelve-vertex polyhedron. The volume of a polyhedron containing an interstitial atom can be increased by decreasing the number of edges. In the case of a deltahedron, this can be done by converting pairs of triangular faces into single quadrilateral faces by rupture of the edge separating the two triangular faces, i.e.,

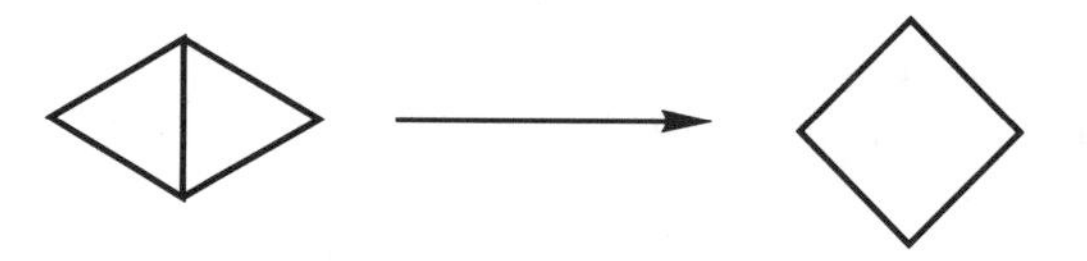

This process corresponds to the "diamond-square" portion of the diamond-square-diamond process involved in polyhedral rearrangements.[28,29,30] For example, rupture of six edges in this manner from an icosahedron can give a cuboctahedron, as indicated in Figure 6–5. An n-vertex non-deltahedron derived from an n-vertex deltahedron by volume expansion through edge-rupture in this manner and containing an interstitial atom may function as a globally delocalized $2n + 2$ skeletal electron system like the original n-vertex

[27]G. Ciani, L. Garlaschelli, A. Sironi, and S. Martinengo, *Chem. Comm.*, 563, 1981.
[28]W. N. Lipscomb, *Science*, **153**, 373, 1966.
[29]R. B. King, *Inorg. Chim. Acta*, **49**, 237, 1981.
[30]R. B. King, *Theor. Chim. Acta*, **64**, 439, 1984.

deltahedron. Such non-deltahedra can conveniently be called *pseudodeltahedra*; they have only triangular and quadrilateral faces, with a limited number of the latter. In an uncentered polyhedron having some faces with more than three edges, such faces may be regarded as holes in the otherwise closed polyhedral surface;[31] such polyhedra correspond to *nido* and *arachno* structures in the boron hydrides (Section 4.4). In a centered pseudodeltahedron, the interstitial atom in the center may be regarded as plugging up the surface holes arising from the non-triangular faces so that globally delocalized bonding is possible.

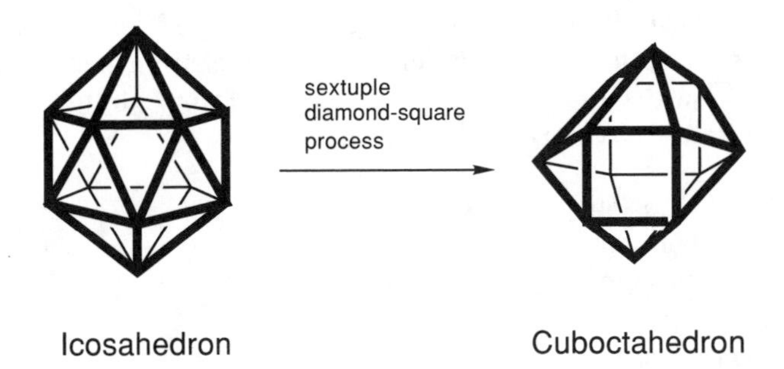

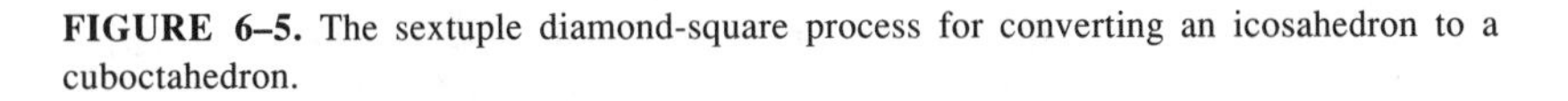

FIGURE 6–5. The sextuple diamond-square process for converting an icosahedron to a cuboctahedron.

The following illustrate electron counting in centered metal carbonyl clusters classified by the polyhedron.

A. OCTAHEDRON (FIGURE 6–1):

A number of octahedral metal carbonyl clusters are known which have a carbon atom in the center of the octahedron. One of the simplest such clusters is $Ru_6C(CO)_{17}$ (ref. 32), which has the 14 skeletal electrons required for a globally delocalized octahedron by the following electron counting scheme:

6 $Ru(CO)_3$ vertices: (6)(2) =	12 electrons
Interstitial carbon atom:	4 electrons
"Deficiency" of one CO group:	–2 electrons
Total skeletal electrons:	*14 electrons*

A related ionic carbon-centered octahedral metal carbonyl cluster is the anionic derivative $Co_6C(CO)_{13}^{2-}$ (ref. 33) which has the 14 skeletal electrons required for a globally delocalized octahedron by the following electron counting scheme:

---

[31]R. B. King, *J. Am. Chem. Soc.*, **94**, 95, 1972.

[32]A. Sirigu, M. Bianchi, and E. Benedetti, *Chem. Comm.*, 596, 1969.

[33]V. G. Albano, D. Braga, and S. Martinengo, *J. Chem. Soc. Dalton*, 981, 1986.

6 normal $Co(CO)_2$ vertices: (6)(1) =	6 electrons
Interstitial carbon atom:	4 electrons
"Extra" CO group:	2 electrons
–2 charge on ion:	2 electrons
Total skeletal electrons:	*14 electrons*

B. TRIGONAL PRISM:

The trigonal prism has six vertices like the octahedron. However, all six vertices of the trigonal prism have degree 3 which match the three internal orbitals of normal vertex atoms suggesting edge-localized skeletal bonding for trigonal prismatic metal carbonyl clusters. A trigonal prismatic cluster of interest is $Co_6C(CO)_{15}^{2-}$ (ref. 34) which has 18 skeletal electrons from the following skeletal electron counting scheme:

6 normal $Co(CO)_2$ vertices: (6)(1) =	6 electrons
Interstitial carbon atom:	4 electrons
3 "extra" carbonyl groups: (3)(2) =	6 electrons
–2 charge on ion:	2 electrons
Total skeletal electrons:	*18 electrons*

These 18 skeletal electrons correspond to two-electron two-center bonds along each of the 9 edges of the Co_6 trigonal prism. The isoelectronic nitrogen-centered Co_6 trigonal prism $Co_6N(CO)_{15}^{-}$ is also known.[35]

Note that an edge-localized Co_6 trigonal prism requires (2)(9) = 18 skeletal electrons for edge-localized bonding along its nine edges, whereas a globally delocalized Co_6 octahedron requires only (2)(6) + 2 = 14 skeletal electrons (= $2n + 2$ for $n = 6$). Therefore, loss of four electrons from a Co_6 trigonal prism can give a Co_6 octahedron. Experimentally, this is achieved by the thermal double decarbonylation of trigonal prismatic $[Et_4N]_2[Co_6C(CO)_{15}]$ to octahedral $[Et_4N]_2[Co_6C(CO)_{13}]$ by heating in boiling tetrahydrofuran. This double decarbonylation of $Co_6C(CO)_{15}^{2-}$ to $Co_6C(CO)_{13}^{2-}$ is a three-dimensional analogue of the triple dehydrogenation of cyclohexane to benzene, with both processes being favored by the resonance stabilization energy arising from the increased bonding delocalization in the product relative to the starting material.

C. TETRAGONAL (SQUARE) ANTIPRISM (FIGURE 6–6):

Both cobalt and nickel form tetragonal antiprismatic metal carbonyl clusters with a carbon atom in the center of the tetragonal antiprism. However, the numbers of skeletal electrons in these two tetragonal

---

[34] S. Martinengo, D. Strumolo, P. Chini, V. G. Albano, and D. Braga, *J. Chem. Soc. Dalton*, 35, 1985.

[35] S. Martinengo, G. Ciani, A. Sironi, B. T. Heaton, and J. Mason, *J. Am. Chem. Soc.*, **101**, 7095, 1979.

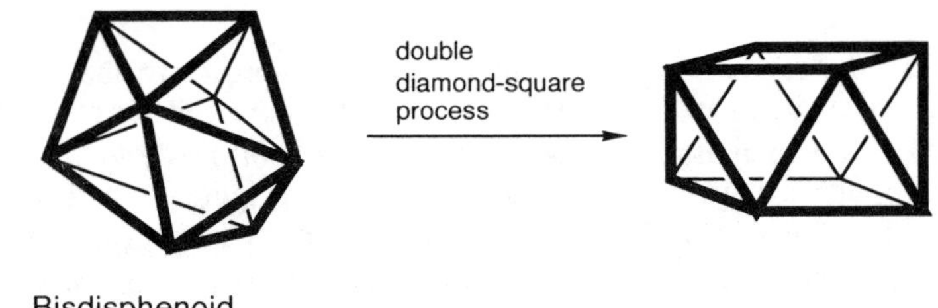

FIGURE 6–6. The double diamond-square process for conversion of a bisdisphenoid to a square antiprism.

antiprismatic complexes differ by four. Thus, the nickel complex $Ni_8C(CO)_{16}^{2-}$ (ref. 36) has 22 skeletal electrons arising from the following electron counting procedure:

8 normal $Ni(CO)_2$ vertices: (8)(2) =	16 electrons
Interstitial carbon atom:	4 electrons
–2 charge on ion:	2 electrons
Total apparent skeletal electrons:	*22 electrons*

These apparent 22 skeletal electrons correspond to the Ni_8 tetragonal antiprism functioning as an *arachno* polyhedron (Section 4.4) with $2n + 6$ skeletal electrons for $n = 8$ in accord with the two tetragonal faces of the tetragonal antiprism. However, the cobalt complex, $Co_8C(CO)_{18}^{2-}$ (ref. 37) has only 18 skeletal electrons arising from the following electron counting procedure:

8 normal $Co(CO)_2$ vertices: (8)(1) =	8 electrons
2 "extra" CO groups: (2)(2) =	4 electrons
Interstitial carbon atom:	4 electrons
–2 charge on ion:	2 electrons
Total apparent skeletal electrons:	*18 electrons*

These apparent 18 skeletal electrons correspond to the Co_8 tetragonal antiprism functioning as an eight-vertex globally delocalized pseudo-deltahedron. This Co_8 polyhedron is thus topologically (although not metrically) a square antiprism with two quadrilateral faces rather than the unique eight-vertex deltahedron, namely the bisdisphenoid (Figure 4–5),

[36] A. Ceriotti, G. Longoni, M. Manassero, M. Perego, and M. Sansoni, *Inorg. Chem.*, **24**, 117, 1985.

[37] V. G. Albano, P. Chini, G. Ciani, S. Martinengo, and M. Sansoni, *J. Chem. Soc. Dalton*, 463, 1978.

which has four degree 4 and four degree 5 vertices. Apparently, the size and shape of the cavity in a Co_8 bisdisphenoid cannot comfortably accommodate an interstitial carbon atom, so that expansion of the bisdisphenoid volume occurs through a double diamond-square process (Figure 6–6). The Co_8 tetragonal antiprism in $Co_8C(CO)_{18}^{2-}$ is thus the globally delocalized pseudodeltahedron which has the smallest number of vertices that has been observed.

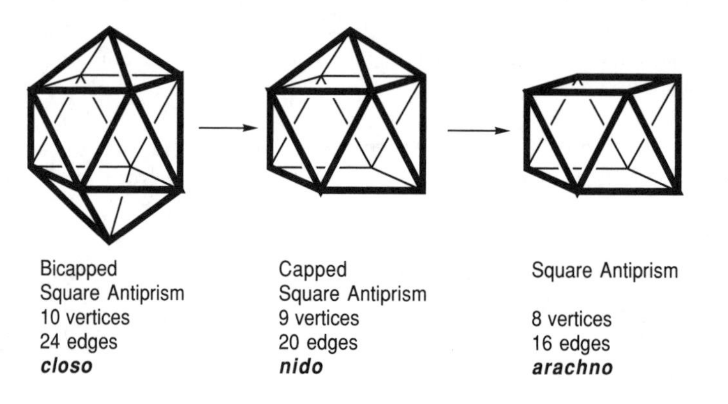

FIGURE 6–7. Conversion of a 10-vertex bicapped square antiprism to a 9-vertex *nido* capped square antiprism and an 8-vertex *arachno* square antiprism.

D. CAPPED SQUARE ANTIPRISM (FIGURE 6–7):

The nickel carbonyl carbide anion $Ni_9C(CO)_{17}^{2-}$ has $18 - 2 + 4 + 2 = 22$ skeletal electrons arising from 9 $Ni(CO)_2$ vertices, a deficiency of one CO group, the interstitial carbon atom, and the –2 charge, respectively. This corresponds to the *nido* $2n + 4$ skeletal electron (for $n = 9$) capped square antiprism with one remaining square face (Figure 6–7).

E. BICAPPED SQUARE ANTIPRISM (FIGURE 6–7):

The nickel carbonyl carbide anion $Ni_{10}C(CO)_{18}^{2-}$ has $20 - 4 + 4 + 2 = 22$ skeletal electrons arising from the 10 $Ni(CO)_2$ vertices, the deficiency of 2 CO groups, the interstitial carbon atom, and the –2 charge, respectively, corresponding to the $2n + 2$ skeletal electrons (for $n = 10$) required for the observed closed deltahedron, namely the D_{4d} bicapped square antiprism (Figure 6–7).

Note that the bicapped square antiprismatic $Ni_{10}C(CO)_{18}^{2-}$, the capped square antiprismatic $Ni_9C(CO)_{17}^{2-}$, and the square antiprismatic $Ni_8C(CO)_{16}^{2-}$ all have 22 skeletal electrons. This represents an excellent illustration of the effects of the successive removal of vertices and associated edges from a closed deltahedron having $2n + 2$ skeletal electrons to give successively a *nido* polyhedron having one quadrilateral face and $2n' + 4$ (where $n' = n - 1$) skeletal electrons and an *arachno* polyhedron having two quadrilateral faces and $2n'' + 6$ (where $n'' = n - 2$) skeletal electrons (Figure 6–7). Note that each vertex removed from the bicapped square antiprism is a degree 4 vertex (i.e., the axial vertices). Removal of such a degree 4 vertex also removes the four edges attached to that vertex.

F. PENTAGONAL ANTIPRISM (FIGURE 6–8):

The pentagonal antiprism is an *arachno* polyhedron obtained by removal of two antipodal vertices from an icosahedron (Figure 6–8). An example of a centered pentagonal antiprismatic complex is the germanium-centered derivative $Ni_{10}Ge(CO)_{20}^{2-}$ (ref. 38), which has 26 skeletal electrons from the following electron-counting scheme:

10 normal $Ni(CO)_2$ vertices: (10)(2) =	20 electrons
Interstitial Ge atom:	4 electrons
–2 charge on ion:	2 electrons
Total apparent skeletal electrons:	*26 electrons*

The observed 26 skeletal electrons is the expected $2n + 6$ for $n = 10$ electron count for an *arachno* 10-vertex polyhedron.

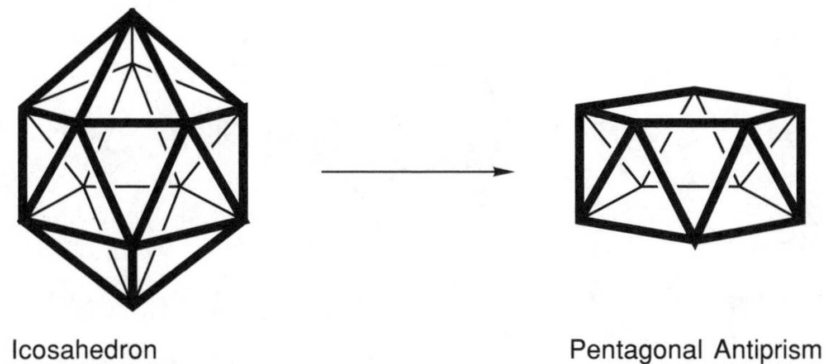

FIGURE 6–8. Conversion of an icosahedron to a pentagonal antiprism by removal of two antipodal vertices and the five edges associated with each antipodal vertex.

G. ICOSAHEDRON (FIGURE 6–8):

Several icosahedral metal carbonyl clusters are known in which an interstitial heavier main group atom, such as germanium, tin or antimony, is located in the center of the icosahedron. These include the following two derivatives, both of which have the 26 skeletal electrons ($= 2n + 2$ for $n = 12$) for a globally delocalized icosahedron:

---

[38] A. Ceriotti, F. Demartin, B. T. Heaton, P. Ingallina, G. Longoni, M. Manassero, M. Marchionna, and N. Masciocchi, *Chem. Comm.*, 786, 1989.

1. $Ni_{12}E(CO)_{22}^{2-}$ (E = Ge, Sn)[38]: 24 – 4 + 4 + 2 = 26 skeletal electrons arising from 12 $Ni(CO)_2$ vertices, a deficit of 2 CO groups, the interstitial Ge or Sn atom, and the –2 charge, respectively;
2. $Rh_{12}Sb(CO)_{27}^{3-}$ (ref. 39): 12 + 6 + 5 + 3 = 26 skeletal electrons arising from 12 $Rh(CO)_2$ vertices, a surplus of 3 CO groups, the interstitial Sb atom, and the –3 charge, respectively.

A more complicated icosahedral metal cluster $Ni_{13}Sb_2(CO)_{24}^{4-}$ (which can be described more precisely as $Ni_{10}[SbNi(CO)_3]_2(Ni)(CO)_{18}^{4-}$)[40] is of interest in having an interstitial *nickel* atom rather than an interstitial antimony atom, such as is found in $Rh_{12}Sb(CO)_{27}^{3-}$. The metal framework of this icosahedral cluster is depicted in Figure 6–9. The required 26 skeletal electrons can be obtained by the following electron counting scheme:

10 normal $Ni(CO)_2$ vertices: (10)(2) =	20 electrons
Deficiency of 2 CO groups: (–2)(2) =	–4 electrons
2 $Sb \rightarrow Ni(CO)_3$ vertices: (2)(5 – 2) =	6 electrons
Interstitial Ni atom:	0 electrons
–4 charge on ion	4 electrons
Total apparent skeletal electrons:	*26 electrons*

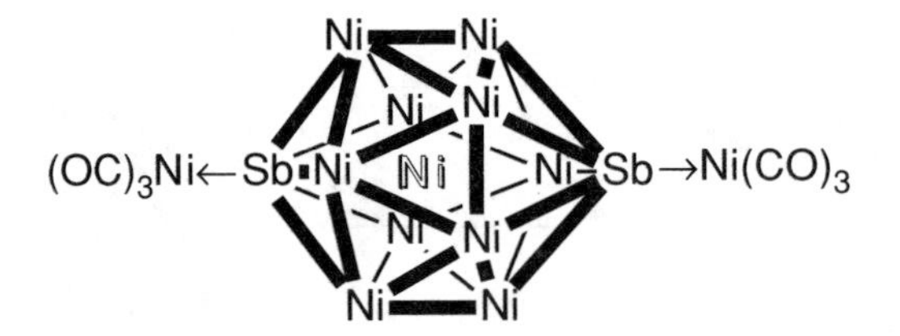

FIGURE 6–9. Metal framework of the icosahedral cluster $Ni_{13}Sb_2(CO)_{24}^{4-}$ = $Ni_{10}[SbNi(CO)_3]_2(Ni)(CO)_{18}^{4-}$.

Note the following features of this model for electron counting in $Ni_{13}Sb_2(CO)_{24}^{4-}$:

1. The two $SbNi(CO)_3$ vertices are donors of three skeletal electrons each, since two of the five valence electrons of the antimony atoms are coordinated to the external $Ni(CO)_3$ groups (Figure 6–9), leaving three electrons for the skeletal bonding.

---

[39] J. L. Vidal and J. M. Troup, *J. Organometal. Chem.*, **213**, 351, 1981.

[40] V. G. Albano, F. Demartin, M. C. Iapalucci, G. Longoni, A. Sironi, and V. Zanotti, *Chem. Comm.*, 547, 1990.

2. The interstitial nickel atom in $Ni_{13}Sb_2(CO)_{24}^{4-}$, which is surrounded by an icosahedron of nickel atoms, is a donor of zero skeletal electrons, since its ten valence electrons are *d* electrons which must go into the quintuply degenerate H_g antibonding level of the Ni_{12} icosahedron, corresponding to the −1 eigenvalue (Figure 4–6). These nickel *d* electrons thus cannot participate in skeletal bonding. Furthermore, the reported[40] oxidation of $Ni_{13}Sb_2(CO)_{24}^{4-}$ to $Ni_{13}Sb_2(CO)_{24}^{3-}$ and $Ni_{13}Sb_2(CO)_{24}^{2-}$ can simply involve removal of one or two of these essentially non-bonding nickel *d* electrons, which thereby affect very little the total energy of the cluster.

H. CUBOCTAHEDRON (FIGURE 6–5):

The icosahedral clusters $Ni_{12}E(CO)_{22}^{2-}$ (E = Ge, Sn), $Ni_{13}Sb_2(CO)_{24}^{4-}$, and $Rh_{12}Sb(CO)_{27}^{3-}$ have structures with an interstitial atom in the center of an Ni_{12} or Rh_{12} icosahedron. In some cases, the volume requirements of an interstitial atom can lead to expansion of an icosahedral cavity to a cuboctahedral cavity by means of the sextuple diamond-square process depicted in Figure 6–5, since the volume of a polyhedron containing an interstitial atom can be increased by decreasing the number of edges. A cuboctahedron can function as a globally delocalized (2)(12) + 2 = 26 skeletal electron 12-vertex pseudodeltahedron. An excellent example of a globally delocalized centered cuboctahedral metal carbonyl cluster is the rhodium carbonyl anion[41] $Rh_{12}(Rh)(CO)_{24}H_3^{2-}$, which has the required 26 skeletal electrons from the following electron counting scheme:

12 $Rh(CO)_2$ vertices: (12)[9 – (4)(2)] =	12 electrons
Interstitial rhodium atom	9 electrons
3 hydrogen atoms: (3)(1) =	3 electrons
–2 charge:	2 electrons
Total skeletal electrons:	*26 electrons*

6.6 METAL CARBONYL CLUSTERS HAVING FUSED OCTAHEDRA

Numerous polycyclic benzenoid hydrocarbons, of which naphthalene and anthracene are familiar examples, can be constructed by edge-sharing fusion of benzene rings (carbon hexagons). Similarly, metal carbonyl cluster polyhedra can be fused by sharing of edges or faces to give three-dimensional analogues of polycyclic aromatic hydrocarbons. The most extensive series of such metal carbonyl derivatives having multiple polyhedral cavities are obtained from

[41] V. G. Albano, A. Ceriotti, P. Chini, G. Ciani, S. Martinengo, and W. M. Anker, *Chem. Comm.* 859, 1975.

metal octahedra; such structures can be viewed as pieces of an infinite three-dimensional bulk metal lattice. Indeed, Teo[42] has shown that the Hume-Rothery rule[43] for electron counting in brasses can be extended to close packed high nuclearity metal clusters. Other aspects of the fusion of cluster polyhedra have been treated by Mingos[44,45] and by Slovokhotov and Struchkov.[46] Figure 6–10 depicts some metal carbonyl clusters constructed by fusion of metal octahedra as analogues of benzenoid aromatic hydrocarbons constructed by fusion of carbon hexagons. The specific systems, which are mainly rhodium carbonyl clusters, are discussed below.

A. BIPHENYL ANALOGUE, $Rh_{12}(CO)_{30}^{2-}$:

The structure of $Rh_{12}(CO)_{30}^{2-}$ (ref. 47) consists of two Rh_6 octahedra joined by a rhodium-rhodium bond analogous to biphenyl, in which two C_6 hexagons (benzene rings) are joined by a carbon-carbon bond. Such a combination of two octahedra requires 28 skeletal electrons, namely 14 for each octahedron ($2n + 2$ where $n = 6$). These 28 skeletal electrons can be obtained as follows:

12 $Rh(CO)_2$ vertices: (12)(1) =	12 electrons
6 "extra" CO groups: (6)(2) =	12 electrons
Rhodium-rhodium bond:	2 electrons
–2 charge on ion:	2 electrons
Total apparent skeletal electrons:	*28 electrons*

B. EDGE-SHARING NAPHTHALENE ANALOGUE, $Ru_{10}C_2(CO)_{24}^{2-}$:

This ruthenium carbonyl cluster[48] consists of two globally delocalized octahedra sharing an edge and with a carbon atom in the center of each octahedron. The two ruthenium atoms of the shared edge (starred in Figure 6–10) use five internal orbitals, whereas the other eight ruthenium atoms use the normal three internal orbitals. An electron counting scheme for $Ru_{10}C_2(CO)_{24}^{2-}$ can be described as follows:

---

[42]B. K. Teo, *Chem. Comm.*, 1362, 1983.

[43]W. Hume-Rothery, *The Metallic State*, Oxford University Press, New York, 1931, 328.

[44]D. M. P. Mingos, *Chem. Comm.*, 706, 1985.

[45]D. M. P. Mingos, *Chem. Comm.*, 1352, 1985.

[46]Yu. L. Slovokhotov and Yu. T. Struchkov, *J. Organometal. Chem.*, **258**, 47, 1983.

[47]V. G. Albano and P. L. Bellon, *J. Organometal. Chem.*, **19**, 405, 1969.

[48]C.-M. T. Hayward, J. R. Shapley, M. R. Churchill, C. Bueno, and A. L. Rheingold, *J. Am. Chem. Soc.*, **104**, 7347, 1982.

(a) Source of skeletal electrons:

8 $Ru(CO)_2$ vertices using 3 internal orbitals:	
(8)[8 – (9–3–2)(2)] =	0 electrons
2 $Ru(CO)_2$ vertices using 5 internal orbitals:	
(2)[8 – (9–5–2)(2)] =	8 electrons
4 "extra" CO groups: (4)(2) =	8 electrons
2 interstitial carbon atoms: (2)(4) =	8 electrons
–2 charge on ion:	2 electrons
Total available skeletal electrons:	*26 electrons*

(b) Use of skeletal electrons:

10 Ru–Ru surface bonds: (10)(2) =	20 electrons
1 Ru–Ru bond along shared edge (starred Ru atoms):	2 electrons
2 6-center core bonds: (2)(2)	4 electrons
Total skeletal electrons required:	*26 electrons*

The published[49] electron-counting scheme for $Ru_{10}C_2(CO)_{24}^{2-}$ is wrong since the –2 charge on the ion was overlooked in that paper.

C. FACE-SHARING NAPHTHALENE ANALOGUE, $Rh_9(CO)_{19}^{3-}$:

The structure of $Rh_9(CO)_{19}^{3-}$ (ref. 50) consists of a pair of octahedra having a (triangular) face in common (Figure 6–10), analogous to naphthalene which consists of two carbon hexagons with an edge in common. The face-shared pair of octahedra has 9 vertices, 21 edges, and 14 faces, like the tricapped trigonal prism (Figure 4–5), which is the nine-vertex deltahedron found in systems having $2n + 2 = 20$ skeletal electrons ($n = 9$), so that a bonding scheme having a K_9 complete graph for the core bonding is reasonable for a face-sharing fused pair of octahedra just as it is for the tricapped trigonal prism. However, in the fused pair of octahedra, the three rhodium vertices common to both octahedra (starred in Figure 6–10) use four internal orbitals, whereas the six rhodium vertices belonging to only one of the octahedra use the normal three internal orbitals. This leads to the following electron counting scheme for $Rh_9(CO)_{19}^{3-}$:

---

[49]R. B. King, *Inorg. Chim. Acta*, **116**, 125, 1986.

[50]S. Martinengo, A. Fumagalli, R. Bonfichi, G. Ciani, and A. Sironi, *Chem. Comm.*, 825, 1982.

(a) Source of skeletal electrons:

6 $Rh(CO)_2$ groups present in only one octahedron and therefore using only 3 internal orbitals: (6)[9–(9–3–2)(2)] =	6 electrons
3 $Rh(CO)_2$ groups common to both octahedra and therefore using 4 internal orbitals: (3)[9–(9–4–2)(2)] =	9 electrons
1 "extra" CO group:	2 electrons
–3 charge on ion:	3 electrons
Total available skeletal electrons:	*20 electrons*

(b) Use of skeletal electrons:

9 Rh–Rh surface bonds: (9)(2) =	18 electrons
1 9-center core bond: (1)(2) =	2 electrons
Total skeletal electrons required:	*20 electrons*

D. ANTHRACENE ANALOGUE, $H_2Rh_{12}(CO)_{25}$:

The structure of $H_2Rh_{12}(CO)_{25}$ (ref. 51) consists of a linear chain of three fused octahedra similar to the fusion of three benzene rings to form anthracene (Figure 6–10). In $H_2Rh_{12}(CO)_{25}$, the distance between the three vertex atoms of the triangular face unique to the octahedron at one end of the chain and the three vertex atoms unique to the octahedron at the other end of the chain is too large for twelve-center core bonding described by a single K_{12} complete graph analogous to the K_9 graph used to describe the nine-center core bonding in the above naphthalene analogue $Rh_9(CO)_{19}^{3-}$. Instead, in $H_2Rh_{12}(CO)_{25}$, the core bonding consists of two complete graphs, one associated with the octahedron at one end of the chain and the other associated with the octahedron at the other end of the chain. This leads to the following electron counting scheme for $H_2Rh_{12}(CO)_{25}$:

(a) Source of skeletal electrons:

6 $Rh(CO)_2$ groups present in only one octahedron and therefore using 3 internal orbitals: (6)[9–(9–3–2)(2)] =	6 electrons
6 $Rh(CO)_2$ groups common to two octahedra (starred in Figure 6–10) and therefore using 4 internal orbitals: (6)[9–(9–4–2)(2)] =	18 electrons
1 "extra" CO group: (1)(2) =	2 electrons
2 hydrogen atoms: (2)(1) =	2 electrons
Total available skeletal electrons:	*28 electrons*

---

[51] G. Ciani, A. Sironi, and S. Martinengo, *Chem. Comm.*, 1757, 1985.

(b) Use of skeletal electrons:

12 Rh–Rh surface bonds: (12)(2) =	24 electrons
2 multicenter core bonds: (2)(2) =	4 electrons
Total skeletal electrons required:	*28 electrons*

The analysis of the bonding topologies in the naphthalene analogue $Rh_9(CO)_{19}^{3-}$ and the anthracene analogue $H_2Rh_{12}(CO)_{25}$ suggests that, in a linear chain of an odd number of face-sharing octahedra, the core bonding occurs in alternate octahedra including the octahedra at both ends, whereas in a linear chain of an even number of face-sharing octahedra, the core bonding consists of a K_9 graph in the two octahedra at one end followed by core bonding in alternate octahedra along the remainder of the chain. Such ideas might ultimately prove relevant in the construction of one-dimensional chains of fused polyhedra having novel metallic properties.

E. PERINAPHTHENE ANALOGUE, $Rh_{11}(CO)_{23}^{3-}$:

The cluster $Rh_{11}(CO)_{23}^{3-}$ (ref. 52) consists of three fused octahedra. The six rhodium atoms unique to a single octahedron are considered to use the normal three internal orbitals, the three rhodium atoms shared by two octahedra are considered to use four internal orbitals, and the two rhodium atoms shared by all three octahedra are considered to use five internal orbitals. Each of the three octahedral cavities contains a K_n multicenter core bond, and in addition there is a "hidden" two-center two-electron bond between the two rhodium vertices common to all three octahedra. This leads to the following electron-counting scheme for $Rh_{11}(CO)_{23}^{3-}$:

(a) Source of skeletal electrons:

6 $Rh(CO)_2$ vertices present in only 1 octahedron and therefore using 3 internal orbitals: (6)[9–(9–3–2)(2)] =	6 electrons
3 $Rh(CO)_2$ vertices common to 2 octahedra and therefore using 4 internal orbitals: (3)[9–(9–4–2)(2)] =	9 electrons
2 $Rh(CO)_2$ vertices common to all 3 octahedra and therefore using 5 internal orbitals: (2)[9–(9–5–2)(2)] =	10 electrons
1 "extra" CO group: (1)(2) =	2 electrons
–3 charge on ion:	3 electrons
Total available skeletal electrons:	*30 electrons*

---

[52]G. Ciani, A. Sironi, and S. Martinengo, *J. Chem. Soc. Dalton*, 519, 1981.

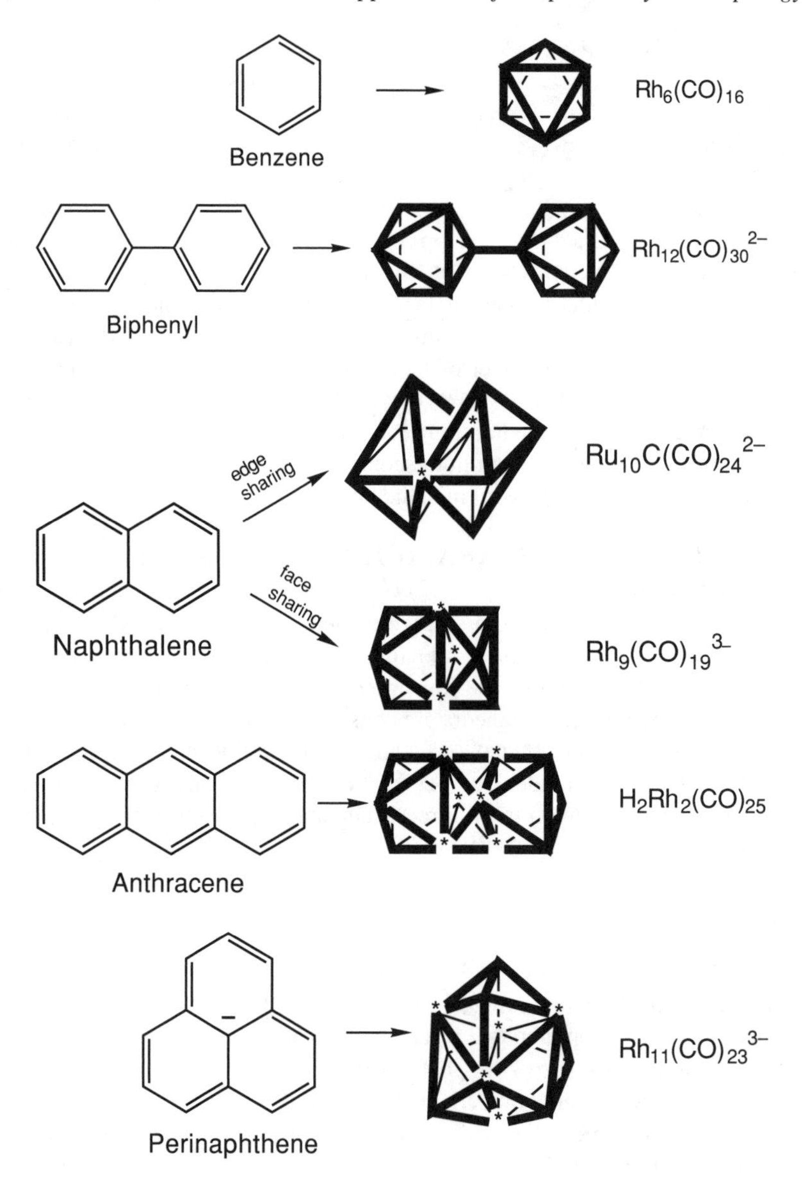

FIGURE 6–10. Analogies between the fusion of metal octahedra in metal carbonyl clusters and the fusion of benzene rings in planar polycyclic aromatic hydrocarbons. Starred (*) vertices are shared by more than one metal octahedron.

(b) Use of skeletal electrons:

11 Rh–Rh surface bonds: (11)(2) =	22 electrons
3 core bonds in the 3 octahedral cavities: (3)(2) =	6 electrons
1 "hidden" two-center two-electron bond between the Rh vertices common to all 3 octahedra:	2 electrons
Total skeletal electrons required:	*30 electrons*

Chapter 7

SOME EARLY TRANSITION METAL AND COINAGE METAL CLUSTERS WITH SPECIAL FEATURES

7.1 EARLY TRANSITION METAL HALIDE CLUSTERS—A COMPARISON OF EDGE-LOCALIZED, FACE-LOCALIZED, AND GLOBALLY DELOCALIZED OCTAHEDRA

Early transition metals such as molybdenum and niobium form discrete metal halide clusters consisting of an M_6 metal octahedron and edge-bridging (μ_2-X) or face-bridging (μ_3-X) halogen atoms. Metal polyhedra other than M_6 octahedra are not generally found in early transition metal halide clusters. The edge-bridging (μ_2-X) halogen atoms as neutral ligands are donors of three electrons each through two electron pairs, whereas the face-bridging (μ_3-X) halogen atoms as neutral ligands are donors of five electrons each through three electron pairs. In the actual three-dimensional structures, the electron pairs of such halogen atoms not required for the primary bridging within a cluster octahedron may be donated to adjacent octahedra in the actual three-dimensional structure. Such lone pairs from external halogen atoms must be considered in electron-counting schemes in order to obtain meaningful electron counts.

The chemical bonding topologies of octahedral early transition metal halide clusters can be divided into the following three types:

1. **Edge-localized:** 12 two-center bonds along the 12 edges of the M_6 octahedron requiring 24 skeletal electrons and 24 internal orbitals corresponding to 4 internal orbitals on each vertex atom.
2. **Face-localized:** 8 three-center bonds in the 8 faces of the M_6 octahedron requiring 16 skeletal electrons and 24 internal orbitals corresponding to 4 internal orbitals for each vertex atom. Face-localized M_6 octahedra thus require 8 skeletal electrons fewer than edge-localized octahedra.
3. **Globally delocalized:** 6 two-center bonds delocalized in the surface of the octahedron and a single six-center bond in the core of the octahedron requiring 14 skeletal electrons and 18 internal orbitals, corresponding to only 3 internal orbitals for each vertex atom. The globally delocalized octahedra found in early transition metal halide clusters are similar to those found in octahedral boranes and carboranes (e.g., $B_6H_6^{2-}$ and $C_2B_4H_6$ discussed in Chapter 4) and octahedral metal carbonyl clusters (e.g., $Rh_6(CO)_{16}$ and $Ni_6(CO)_{12}^{2-}$ discussed in Chapter 6).

The prototypical examples of edge-localized octahedral early transition metal halide clusters are the molybdenum(II) halide derivatives generically represented as $Mo_6X_8L_6^{4+}$, including "molybdenum dichloride", $Mo_6(\mu_3\text{-}Cl)_8Cl_2Cl_{4/2}$.[1] The structures of these compounds consist of Mo_6 octahedra, a face-bridging halogen atom (μ_3-X) in each of the eight octahedral faces, and one bond from each molybdenum vertex to an external ligand (L) which may be a halogen atom bridging to another Mo_6 octahedron. The resulting geometry contains an Mo_6 octahedron inside an X_8 cube (Figure 7–1a). The coordination polyhedron of each of the vertex molybdenum atoms is a capped square antiprism (Figure 7–1b) with the external ligand in the axial position (*a* in Figure 7–1b), the four bonds to face-bridging halogen atoms in the four medial positions (*m* in Figure 7–1b), and the four internal orbitals in the basal positions (*b* in Figure 7–1b), forming the two-center bonds to adjacent molybdenum atoms. An L–Mo vertex using four internal orbitals and thus five external orbitals is a (5)(2) – 6 – 2 = 2 electron *acceptor* (or –2 electron donor) after allowing for six electrons from the neutral molybdenum atom and two electrons from the neutral ligand L. This leads to the following electron counting scheme for $Mo_6X_8L_6^{4+}$:

6 L–Mo vertices: (6)(–2) =	–12 electrons
8 μ_3-X bridges: (8)(5) =	40 electrons
+4 charge:	–4 electrons
Net skeletal electrons:	*24 electrons*

These 24 skeletal electrons are exactly the number required for an edge-localized octahedron having 12 two-center edge bonds as discussed above.

Now consider the octahedral niobium halide clusters of the general type $Nb_6X_{12}L_6^{2+}$, including the binary halide $Nb_6(\mu_2\text{-}Cl)_{12}Cl_{6/3}$ (= Nb_6Cl_{14}).[2] The structures of these compounds consist of Nb_6 octahedra, an edge-bridging halogen atom (μ_2-X) across each of the 12 octahedral edges, and one bond from each niobium vertex to an external ligand (L), which may be a halogen atom bridging to another Nb_6 octahedron such as in Nb_6Cl_{14}. Again, the coordination polyhedron of each of the vertex niobium atoms is a capped square antiprism (Figure 7–1) with the external ligand in the axial position (*a* in Figure 7–1b), four bonds from the medial positions (*m* in Figure 7–1b) to edge-bridging halogen atoms, and four internal orbitals in the basal positions (*b* in Figure 7–1b) to form the three-center face bonds with adjacent niobium atoms. An L–Nb vertex using four internal orbitals and thus five external orbitals is a (5)(2) – 5 – 2 = 3 skeletal electron *acceptor* (or a –3 electron donor), after allowing for five electrons from the neutral niobium atom and

[1]H. Schäfer, H. Schnering, J. Tillack, F. Kuhnen, H. Wöhler, and H. Baumann, Z. *anorg. allgem. Chem.*, **353**, 281, 1965.

[2]A. Simon, H. G. Schnering, H. Wöhler, and H. Schäfer, Z. *anorg. allgem. Chem.*, **339**, 155, 1965.

two electrons from the neutral ligand L. This leads to the following electron-counting scheme for $Nb_6X_{12}L_6^{2+}$:

6 LNb vertices: (6)(–3) =	–18 electrons
12 μ_2-X vertices: (12)(3) =	36 electrons
+2 charge	–2 electrons
Net skeletal electrons:	*16 electrons*

These 16 skeletal electrons are exactly the number required for a face-localized octahedron with its eight three-center face bonds as discussed above.

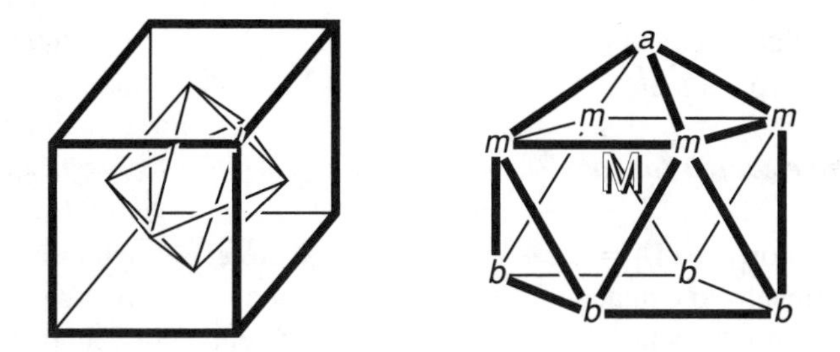

FIGURE 7–1. (a) An M_6 octahedron within an X_8 cube; (b) the capped square antiprism metal (M) coordination in octahedral early transition metal halide clusters.

The face-bridged edge-localized $Mo_6X_8L_6^{4+}$ clusters and the edge-bridged face-localized $Nb_6X_{12}L_6^{2+}$ clusters may be regarded as complementary, since in $Mo_6X_8L_6^{4+}$ the halogen atoms occupy faces and the metal-metal bonds occupy edges, whereas in $Nb_6X_{12}L_6^{2+}$ the roles of the edges and faces are reversed, so that the halogen atoms occupy edges and the metal-metal-metal (three-center) bonds occupy faces. These clusters are discussed in further detail by Johnston and Mingos[3] within the framework of tensor surface harmonic theory (Section 4.5).

There are also a number of discrete octahedral early transition metal halide clusters in which an interstitial atom is located in the center of the transition metal octahedron. An important class of such compounds is constructed from a Zr_6X_{12} unit, consisting of a nitrogen-centered Zr_6 octahedron in which each of its 12 edges is bridged by a halogen atom.[4,5,6,7] In such clusters, each zirconium vertex is bonded to five halogen atoms. Four of these halogen atoms are those bridging edges to a neighboring zirconium atom in the Zr_6 octahedron, whereas the fifth halogen atom bridges from another Zr_6 octahedron. This latter (external) halogen atom may be formally treated as a

---

[3]R. L. Johnston and D. M. P. Mingos, *Inorg. Chem.*, **25**, 1661, 1986.
[4]R. P. Ziebarth and J. D. Corbett, *J. Am. Chem. Soc.*, **107**, 4571, 1985.
[5]R. P. Ziebarth and J. D. Corbett, *J. Am. Chem. Soc.*, **110**, 1132, 1988.
[6]J. D. Smith and J. D. Corbett, *J. Am. Chem. Soc.*, **108**, 1927, 1986.
[7]G. Rosenthal and J. D. Corbett, *Inorg. Chem.*, **27**, 53, 1988.

ligand (L) so that NZr_6Cl_{15}, for example, may be treated as $NZr_6Cl_{12}L_6^{3+}$, in which three of the ligands for each Zr_6 octahedron are three added Cl^-, corresponding to a Cl/Zr ratio of 15/6 rather than the Cl/Zr ratio of 12/6 for the edge-bridged Zr_6Cl_{12} octahedron alone. The remaining three ligands bonded to Zr vertices of the Zr_6 octahedron in NZr_6Cl_{15} are lone pairs from chlorine atoms in adjacent Zr_6Cl_{12} octahedra. The LZr vertices clearly use five external orbitals for bonding to chlorine atoms, making them $(5)(2) - 4 - 2 = 4$ electron *acceptors*, i.e., –4 electron donors. This leads to the following electron-counting scheme for $NZr_6Cl_{12}L_6^{3+}$ (= $NZr_6(\mu_2\text{-}Cl)_{12}Cl_{n-12}$):

6 L–Zr vertices: (6)(–4) =	–24 electrons
12 μ_2-Cl bridges: (12)(3) =	36 electrons
Interstitial nitrogen atom:	5 electrons
+3 charge:	–3 electrons
Net skeletal electrons for each Zr_6 octahedron:	*14 electrons*

This is the correct ($14 = 2n + 2$ for $n = 6$) skeletal electron count for a globally delocalized octahedron (Section 4.3), which has six two-center bonds delocalized into the octahedral surface and one six-center core bond. Note that this bonding topology requires only the normal three internal orbitals from each vertex atom, meaning that each vertex zirconium atom has a manifold of only eight bonding orbitals, namely five external orbitals to chlorine atoms and three rather than four internal orbitals. This corresponds to a 16-electron zirconium configuration similar to that found in a variety of stable zirconium compounds such as $(C_5H_5)_2ZrCl_2$. Also note that the presence of an interstitial atom in the center of an octahedral early transition metal halide cluster favors a globally delocalized chemical bonding topology rather than the edge-localized chemical bonding topology of the uncentered molybdenum halide octahedra $Mo_6X_8L_6^{4+}$ or the face-localized chemical bonding topology in the uncentered niobium halide octahedra.

A feature of centered metal clusters containing interstitial non-transition metal atoms such as Be, B, C, N, Si, Ge or P, which use an sp^3 chemical bonding manifold, is the involvement of all of the valence electrons of the interstitial atom in the skeletal bonding. Thus, the interstitial nitrogen atom in NZr_6Cl_{15} contributes all five of its valence electrons to the skeletal bonding as noted above. However, the situation is more complicated if the interstitial atom is a transition metal such as Cr, Mn, Fe, or Co, found in centered octahedral zirconium cluster iodides.[8] In this case not all of the valence *d* electrons of the interstitial transition metal can participate in the skeletal bonding.

This point is illustrated in Table 7–1. The single interstitial atom *s* orbital and all three of its *p* orbitals (x,y,z) belong to the A_{1g} and T_{1u} irreducible

---

[8]T. Hughbanks, G. Rosenthal, and J. D. Corbett, *J. Am. Chem. Soc.*, **110**, 1511, 1988.

representations, respectively, corresponding to bonding core and surface orbitals of the octahedron, respectively, designated as S^σ and P^σ, respectively, in tensor surface harmonic theory (Figure 4–6 and Section 4.5). For this reason, all of the valence electrons of interstitial non-transition elements using only *s* and *p* orbitals are available for skeletal bonding. However, only three of the five *d* orbitals (*xy*, *xz*, *yz*) of interstitial transition metals belong to the irreducible representations of bonding skeletal orbitals, namely the t_{2g} or P^σ surface orbitals. The remaining two *d* orbitals (z^2, x^2–y^2) belong to the E_g irreducible representation, which appears only in the e_g core *antibonding orbitals* (D^σ orbitals in tensor surface harmonic theory). Valence electrons in the z^2 and x^2–y^2 orbitals of an interstitial transition metal are therefore unavailable for skeletal bonding in a surrounding M_6 metal octahedron but remain as non-bonding electrons in molecular orbital models. Such non-bonding electrons of an interstitial atom can be called *latent* electrons. Because of the unavailability of four valence electrons of an interstitial transition metal for skeletal bonding in a surrounding M_6 metal octahedron, interstitial Cr, Mn, Fe, and Co atoms inside M_6 octahedra are donors of 2, 3, 4, and 5 skeletal electrons, respectively. Thus, the diamagnetic $CrZr_6I_{12}$ can be shown as follows to have the 14 skeletal electrons of a globally delocalized octahedron:

6 L–Zr vertices: (6)(–4) =	–24 electrons
12 μ_2–I bridges: (12)(3) =	36 electrons
Interstitial Cr atom: 6 – 4 =	2 electrons
Net skeletal electrons for each Zr_6 octahedron:	*14 electrons*

TABLE 7–1
Skeletal Molecular Orbitals and Interstitial Atom Orbitals in Centered Octahedral Metal Clusters

Skeletal Molecular Orbitals	Corresponding Interstitial Atom Orbitals
Bonding Orbitals	
a_{1g} core	*s*
t_{1u} surface	*p* orbitals (*x*, *y*, *z*)
t_{2g} surface	*d* orbitals (*xy*, *xz*, *yz*)
Antibonding Orbitals	
t_{2u} surface	no *s*, *p*, or *d* orbitals
t_{1g} surface	no *s*, *p*, or *d* orbitals
t_{1u} core	*p* orbitals (*x*, *y*, *z*)
e_g core	*d* orbitals (z^2, x^2–y^2)

7.2 CENTERED GOLD CLUSTERS—NON-SPHERICAL VALENCE ORBITAL MANIFOLDS

Among the three coinage metals, namely copper, silver, and gold, the most extensive series of clusters are formed by gold.[9,10,11,12,13,14,15,16] Gold clusters are particularly interesting, since gold is an excellent example of a heavy element for which relativistic effects[17] appear to be important in chemical bonding, and gold clusters contain topological features not found in clusters of other transition metals.

The accessible *spd* manifold of most transition metals consists of nine orbitals (sp^3d^5) and has spherical (isotropic) geometry extending equally in all three dimensions, where the geometry of an orbital manifold relates to contours of the sum $\Sigma\psi^2$ (Section 3.1) over all orbitals in the manifold. Filling this accessible spherical *spd* manifold with electrons from either the central metal atom or its surrounding ligands results in the familiar 18-electron configuration of the next rare gas.

A specific feature of the chemical bonding in some systems containing the late transition metals observed by Nyholm[18] as early as 1961 is the shifting of one or two of the outer *p* orbitals to such high energies that they no longer participate in the chemical bonding. If one *p* orbital is so shifted to become antibonding, then the accessible *spd* orbital manifold contains only eight orbitals (sp^2d^5) and has the geometry of a torus or doughnut. The "missing" *p* orbital is responsible for the hole in the doughnut. This toroidal sp^2d^5 manifold can bond only in the two dimensions of the plane of the ring of the torus. Filling this sp^2d^5 manifold of eight orbitals with electrons leads to the 16-electron configuration found in square planar complexes of the d^8 transition metals such as Rh(I), Ir(I), Ni(II), Pd(II), Pt(II), and Au(III). The locations of the four ligands in these square planar complexes can be considered to be points on the surface of the torus corresponding to the sp^2d^5 manifold.

In some structures containing the late 5*d* transition and post-transition metals Pt, Au, Hg, and Tl, two of the outer *p* orbitals are raised to

---

[9]H. Schmidbaur and K. C. Dash, *Adv. Inorg. Chem. Radiochem.*, **25**, 243, 1982.

[10]J. J. Stegerda, J. J. Bour, and J. W. A. van der Velden, *Rec. Trav. Chim. Pays-Bas*, **101**, 164, 1982.

[11]D. M. P. Mingos, *J. Chem. Soc. Dalton*, 1163, 1976.

[12]D. M. P. Mingos, *Philos. Trans. R. Soc. Lond. A*, **308**, 75, 1982.

[13]D. G. Evans and D. M. P. Mingos, *J. Organometal. Chem.*, **232**, 171, 1982.

[14]D. M. P. Mingos, *Polyhedron*, **3**, 1289, 1984.

[15]K. P. Hall, D. I. Gilmour, and D. M. P. Mingos, *J. Organometal. Chem.*, **268**, 275, 1984.

[16]Z. Lin, R. P. F. Kanters, and D. M. P. Mingos, *Inorg. Chem.*, **30**, 91, 1991.

[17]P. Pyykkö and J.-P. Desclaux, *Acc. Chem. Res.*, **12**, 276, 1979.

[18]R. S. Nyholm, *Proc. Chem. Soc.*, 273, 1961.

antibonding energy levels. This leaves only one *p* orbital in the accessible *spd* orbital manifold, which now contains 7 orbitals (spd^5) and has cylindrical geometry extending in one axial dimension much further than in the remaining two dimensions. Filling this spd^5 manifold with electrons leads to the 14-electron configuration found in two-coordinate linear complexes of d^{10} metals such as Pt(0), Ag(I), Au(I), Hg(II), and Tl(III). The raising of one or particularly two outer *p* orbitals to antibonding levels has been attributed to relativistic effects.

Thus, in an initial approximation, the spd^5 orbital manifold of an L→Au or X—Au vertex (L = tertiary phosphine or isocyanide ligand; X = halogen or pseudohalogen) in a polyhedral gold cluster may be regarded as having a pair of linear *sp* hybrids. One of these hybrids, corresponding to the unique internal orbital in the previously discussed bonding models (Section 4.2), points towards the center of the polyhedron and thus can participate in core bonding. The other *sp* hybrid corresponds to the external orbital in the previously discussed bonding models (Section 4–2) and overlaps with the bonding orbital from the external L or X ligand. In this initial approximation, the five *d* orbitals of the gold vertex are essentially non-bonding and are filled with electron pairs, thereby using 10 of the 11 valence electrons of a neutral gold atom. As a result of this, the L→Au and X—Au vertices are donors of one and zero skeletal electrons, respectively. In this initial approximation, an L→Au vertex functions much like a large hydrogen atom. Thus the binuclear derivative[19] Ph_3P→Au—Au←PPh_3 may be regarded as an analogue of dihydrogen, H—H, as well as the "mercurous" (mercury(I)) halides, X—Hg—Hg—X, with which it is isoelectronic.

An important difference between an L→Au vertex and a hydrogen atom is the two empty orthogonal *p* orbitals of the L→Au vertex, namely the *p* orbitals that are not used for the *sp* hybrid mentioned above. These are the *p* orbitals which are raised to antibonding levels as noted above so that they are not contained in the spd^5 cylindrical manifold of bonding orbitals. These empty *p* orbitals correspond to the twin internal orbitals in the previously discussed bonding models (Section 4.2) and, although lacking electrons in the initial approximation, are appropriately oriented for surface bonding. Such surface bonding in these gold clusters can involve overlap between a filled *d* orbital of a gold vertex and an empty *p* orbital of an adjacent gold vertex and thus can be viewed as an unusual example of dσ→pσ* or dπ→pπ* bonding, depending on the symmetry of the overlap. Such bonding has been suggested by Dedieu and Hoffmann[20] for closely related Pt(0)–Pt(0) dimers on the basis of extended Hückel calculations. This type of surface bonding, like, for example, the dπ→pπ* backbonding in metal carbonyls, does not affect the electron bookkeeping in the gold cluster but accounts for the bonding rather than non-bonding distances between adjacent gold vertices in gold clusters.

---

[19] D. M. P. Mingos, *Pure Appl. Chem.*, **52**, 705, 1980.

[20] A. Dedieu and R. Hoffmann, *J. Am. Chem. Soc.*, **100**, 2074, 1978.

An important class of gold clusters is the *centered* gold clusters. Such clusters containing n gold atoms consist of a central gold atom surrounded by $n - 1$ peripheral gold atoms. The peripheral gold atoms all have a cylindrical spd^5 manifold of bonding orbitals and can be divided into the following two types:

1. Belt gold atoms, which form a puckered hexagonal or octagonal belt around the center gold atom;
2. Distal gold atoms, which appear above or below the belt gold atoms.

The topology of the centered gold clusters can be considered to be either spherical or toroidal, depending on whether the center gold atom uses a spherical sp^3d^5 manifold or a toroidal sp^2d^5 manifold of bonding orbitals; this distinction between spherical and toroidal centered gold clusters was first recognized by Mingos and coworkers.[21] Centered gold clusters have also been described as "porcupine compounds", since the central gold atom corresponds to the body of the porcupine and the peripheral gold atoms (with cylindrical geometry as noted above) correspond to the quills of the porcupine.[22]

The following features of centered gold clusters make their systematics very different from those of other metal clusters compounds such as the metal carbonyl clusters discussed in Chapter 6:

1. The volume enclosed by the peripheral gold atoms must be large enough to contain the center gold atom. Thus, the volume of a cube of eight peripheral gold atoms is not large enough to contain a ninth central gold atom without some distortion. Therefore, centered cube gold clusters of the stoichiometry $Au_9L_8^+$, such as $Au_9(PPh_3)_8^+$ (ref. 23), are distorted from the ideal O_h symmetry to lower symmetry such as D_3. However, the volume of an icosahedron of 12 peripheral gold atoms is large enough to contain a thirteenth central gold atom without any distortion. The peripheral gold polyhedron of a spherical gold cluster containing fewer than 13 total gold atoms is generally based on an undistorted icosahedral fragment which has a large enough volume for the center gold atom.
2. The overlap topology at the core of a centered Au_n cluster from the $n - 1$ unique internal orbitals of the peripheral gold atoms is not that of a K_{n-1} complete graph as in other globally delocalized metal clusters (Chapter 4). Instead, the overlap topology of the unique internal orbitals of the peripheral gold atoms corresponds to the polyhedron formed by the peripheral gold atoms. This can be related to the sharper "focus" of

---

[21] C. E. Briant, K. P. Hall, A. C. Wheeler, and D. M. P. Mingos, *Chem. Comm.*, 248, 1984.

[22] R. B. King, *Prog. Inorg. Chem.*, **15**, 287, 1972.

[23] J. G. M. van der Linden, M. L. H. Paulissen, and J. E. J. Schmitz, *J. Am. Chem. Soc.*, **105**, 1903, 1983.

the cylindrical seven-orbital spd^5 manifold of the peripheral gold atoms relative to that of the spherical four-orbital sp^3 and nine-orbital sp^3d^5 manifolds of the clusters discussed in Chapters 4 and 6. Thus, the number of positive eigenvalues of the graphs corresponding to the peripheral gold polyhedra relates to the number of bonding orbitals in the centered gold clusters based on the relation between graph theory and Hückel theory outlined in Section 4.1.

3. The center gold atom has 11 valence electrons. All but one of these electrons are needed to fill its five *d* orbitals. The remaining electron is in the spherically symmetric *s* orbital, which is the orbital of the center gold atom overlapping with the unique internal orbitals of the cylindrical spd^5 manifold of the peripheral gold atoms. This overlap lowers the energy of the lowest (fully symmetric) cluster bonding orbital without adding any new bonding orbitals. The center gold atom is therefore a donor of one skeletal electron.
4. Mingos and coworkers[21] have observed a $12p + 16$ electron rule for toroidal centered gold clusters and a $12p + 18$ electron rule for spherical centered gold clusters where $p = n - 1$ is the number of peripheral gold atoms. These numbers count not only the skeletal electrons but also the 10 electrons needed to fill the five *d* orbitals of each peripheral gold atom and the 2 electrons needed for one bond from each peripheral gold atom to an external L or X group. The $12p$ terms in Mingos' total electron numbers thus correspond to non-skeletal electrons involving only the peripheral gold atoms, leaving 16 or 18 electrons for a center gold atom with toroidal or spherical geometry, respectively. This corresponds exactly to the number of electrons required to fill the eight-orbital toroidal sp^2d^5 manifold or the nine-orbital spherical sp^3d^5 manifold, respectively. Subtracting 10 from the 16 or 18 electrons allocated to the center gold atom for its five *d* orbitals leaves 6 or 8 skeletal electrons for toroidal or spherical gold clusters, respectively.
5. Consider L to be a two-electron donor ligand (e.g., tertiary phosphines or isocyanides) and X to be a one-electron donor ligand (e.g., halogen, pseudohalogen, $Co(CO)_4$, etc.). Then the above considerations give centered toroidal clusters (Figure 7–2) the general formula $Au_nL_yX_{n-1-y}^{(y-5)+}$ and centered spherical clusters (Figure 7–3) the general formula $Au_nL_yX_{n-1-y}^{(y-7)+}$.

All of the centered toroidal gold clusters depicted schematically in Figure 7–2 conform to the $Au_nL_yX_{n-1-y}^{(y-5)+}$ formula mentioned above. Thus the clusters $Au_7L_6^+$, $Au_8(PPh_3)_7^{2+}$, $Au_9(PPh_3)_8^{3+}$, $Au_9(SCN)_3(PCx_3)_5$, and $Au_{10}Cl_3(PCx_2Ph)_6^+$ correspond to this formula with $n = 7$, $y = 6$; $n = 8$, $y = 7$; $n = 9$, $y = 8$; $n = 9$, $y = 5$; and $n = 10$, $y = 6$, respectively. The D_{2h}

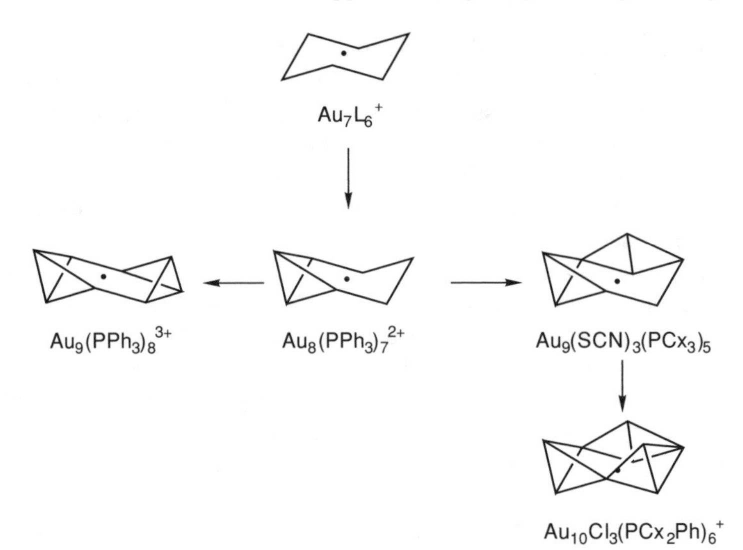

FIGURE 7–2. The structures of some toroidal gold clusters of the general formula $Au_nL_yX_{n-1-y}^{(y-5)+}$. The center gold atoms are shown by a dot (•).

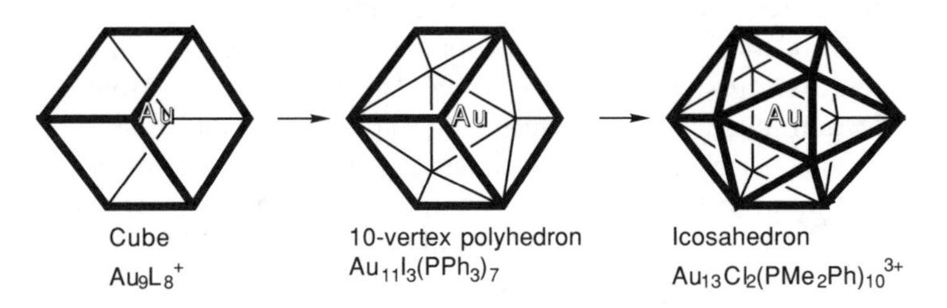

FIGURE 7–3. The structures of some spherical gold clusters of the general formula $Au_nL_yX_{n-1-y}^{(y-7)+}$. The center gold atoms are shown as **Au** .

hexagonal belt cluster $Au_9(PPh_3)_8^{3+}$ and the C_{2v} hexagonal belt cluster $Au_9(SCN)_3(PCx_3)_5$ differ in the placement of the two distal gold atoms relative to the hexagonal belt, as depicted in Figure 7–2. The spectra of the graphs representing the interactions of the peripheral gold atoms in all of the toroidal clusters in Figure 7–2 have three positive eigenvalues, in accord with the presence of three skeletal bonding orbitals corresponding to six skeletal electrons. Adding these six skeletal electrons to the 10 electrons required to fill the five *d* orbitals of the center gold atom gives the 16 electrons required to fill the eight-orbital sp^2d^5 manifold of the bonding orbitals of the center gold atom.

The three spherical gold clusters, $Au_{13}Cl_2(PMe_2Ph)_{10}^{3+}$, $Au_{11}I_3(PPh_3)_7$, and $Au_9(PPh_3)_8^+$, depicted in Figure 7–3 all conform to the $Au_nL_yX_{n-1-y}^{(y-7)+}$ formula noted above with $n = 13$, $y = 10$; $n = 11$, $y = 7$; and $n = 9$, $y = 8$, respectively. The spectra of the graphs representing their peripheral gold atoms all have four positive eigenvalues, in accord with the presence of four skeletal bonding orbitals corresponding to eight skeletal electrons. Addition of these eight skeletal electrons to the 10 electrons required to fill the five *d* orbitals of the center gold atom gives the 18 electrons required to fill the nine-orbital spherical sp^3d^5 manifold of the bonding orbitals of the center gold atom. The peripheral 10-vertex gold polyhedron in $Au_{11}I_3(PPh_3)_7$ can be formed from the peripheral gold icosahedron in $Au_{13}Cl_2(PMe_2Ph)_{10}^{3+}$ by the following two-step process:

1. Removal of a triangular face, including its three vertices, its three edges, and the nine edges connecting this face with the remainder of the icosahedron.
2. Addition of a new vertex in the location of the midpoint of the face that was removed followed by addition of three new edges to connect this new vertex to the degree 3 vertices of the nine-vertex icosahedral fragment produced in the first step.

These processes preserve a C_3 axis of the icosahedron. Also, application of this two-step process twice to an icosahedron so as to preserve a C_3 axis throughout the whole sequence of steps (Figure 7–3) leads to a cube such as the (distorted) cube found in $Au_9(PPh_3)_8^+$ (ref. 23). In this sense, the 10-vertex peripheral gold polyhedron of $Au_{11}I_3(PPh_3)_7$ can be considered to be halfway between an icosahedron and a cube.

Chapter 8

POST-TRANSITION METAL CLUSTERS

8.1 MOLECULAR AND IONIC BARE POST-TRANSITION METAL CLUSTERS

The polyhedral boranes and carboranes discussed in Chapter 4 may be regarded as boron clusters in which the single external orbital of each vertex atom is used for bonding to an external monovalent group. Analogously, the six external orbitals of each transition metal vertex atom in the metal carbonyl clusters discussed in Chapter 6 are used for bonding to ligands such as carbonyl groups, tertiary phosphines, or planar hydrocarbon rings such as cyclopentadienyl. The single external orbital on each gold atom in the centered gold clusters discussed in Section 7.2 is used for bonding to a single tertiary phosphine, halide, or pseudohalide ligand. Electron-richer post-transition metals to the right of gold in the Periodic Table form clusters having *no* external ligands bonded to the vertex atoms. Such metal clusters are conveniently called *bare* metal clusters. Anionic bare metal clusters were first observed by Zintl and coworkers in the 1930s,[1,2,3,4] who obtained the first evidence for anionic clusters of post-transition metals such as tin, lead, antimony, and bismuth through potentiometric titrations with alkali metals in liquid ammonia; for this reason such anionic post-transition metal clusters are often called *Zintl phases*. However, extensive structural information on these anionic post-transition metal clusters was obtained only in the 1970s by Corbett and co-workers,[5] who discovered that complexation with 2,2,2-crypt gave crystals of alkali metal derivatives of such clusters suitable for structure determination by X-ray diffraction. Somewhat earlier, Corbett and co-workers[6] also used X-ray crystallography to obtain definitive structural information on cationic post-transition metal clusters obtained as halometalate salts, particularly $AlCl_4^-$ salts, from highly acidic melts. In recent years, such work on condensed-phase ionic post-transition metal clusters has been complemented by studies on gas phase neutral post-transition metal clusters generated in molecular beam equipment using resistive heating[7,8,9,10] or laser

[1] E. Zintl, J. Goubeau, and W. Dullenkopf, *Z. Phys. Chem., Abt. A*, **154**, 1, 1931.
[2] E. Zintl and A. Harder, *Z. Phys. Chem., Abt. A*, **154**, 47, 1931.
[3] E. Zintl and W. Dullekopf, *Z. Phys. Chem., Abt. B*, **16**, 183, 1932.
[4] E. Zintl and H. Kaiser, *Z. anorg. allgem. Chem.*, **211**, 113, 1933.
[5] J. D. Corbett, *Chem. Rev.*, **85**, 383, 1985.
[6] J. D. Corbett, *Prog. Inorg. Chem.*, **21**, 129, 1976.
[7] K. Sattler, J. Mühlbach, and E. Recknagel, *Phys. Rev. Lett.*, **45**, 821, 1980.
[8] J. Mühlbach, K. Sattler, P. Pfau, and E. Recknagel, *Phys. Lett. A*, **87A**, 415, 1982.
[9] K. Sattler, J. Mühlbach, P. Pfau, and E. Recknagel, *Phys. Lett. A*, **87A**, 418, 1982.
[10] D. Schild, R. Pflaum, K. Sattler, and E. Recknagel, *J. Phys. Chem.*, **91**, 2649, 1987.

vaporization.[11,12,13,14,15,16] Stoichiometries and relative abundances of the gas phase metal clusters can be measured by mass spectrometry. However, definitive experimental methods for the determination of structures of the gas phase post-transition metal clusters analogous to X-ray diffraction for the crystalline ionic post-transition metal clusters are not yet available.

The rules for counting the number of skeletal electrons contributed by each vertex atom can be adapted to vertices consisting of post-transition metals lacking external groups through the following considerations:

1. The post-transition metals under consideration use a nine-orbital sp^3d^5 valence orbital manifold.
2. The post-transition metals (in their zero formal oxidation states) have a total of $10 + G$ valence electrons where G is the highest possible oxidation state of the post-transition metal. G also corresponds to the number of the group in the old version of the Periodic Table, where the post-transition metal in question is located. Thus germanium, tin, and lead have $10 + 4 = 14$ valence electrons; arsenic, antimony, and bismuth have $10 + 5 = 15$ valence electrons; and selenium and tellurium have $10 + 6 = 16$ valence electrons. These numbers of valence electrons for the post-transition metals also correspond to their group numbers in the recently adopted version of the Periodic Table.
3. If the clusters have either two- or three-dimensional aromaticity, as discussed in Section 4.3, three orbitals of each bare metal vertex atom will be required for the internal orbitals (two twin internal orbitals and one unique internal orbital). This leaves $9 - 3 = 6$ external orbitals. Each external orbital of the bare metal vertex atom must be filled with an electron pair, thereby consuming $(2)(6) = 12$ electrons from each bare metal vertex atom.
4. As a result of these considerations, the number of skeletal electrons contributed by each bare metal vertex atom through its three internal orbitals must be $10 + G - 12 = G - 2$.

Application of this procedure to the post-transition metals forming clusters indicates that gallium, indium, and thallium contribute one skeletal electron; germanium, tin, and lead contribute two skeletal electrons; arsenic, antimony,

[11] R. G. Wheeler and M. A. Duncan, *Chem. Phys. Lett.*, **131**, 8, 1986.

[12] R. G. Wheeler, K. LaiHing, W. L. Wilson, J. D. Allen, R. B. King and M. A. Duncan, *J. Am. Chem. Soc.*, **108**, 8101, 1986.

[13] K. LaiHing, R. G. Wheeler, W. L. Wilson, and M. A. Duncan, *J. Chem. Phys.*, **87**, 3401, 1987.

[14] M. E. Geusic, R. R. Freeman, and M. A. Duncan, *J. Chem. Phys.*, **88**, 163, 1988.

[15] M. E. Geusic, R. R. Freeman, and M. A. Duncan, *J. Chem. Phys.*, **89**, 223, 1988.

[16] R. G. Wheeler, K. LaiHing, W. L. Wilson, and M. A. Duncan, *J. Chem. Phys.*, **88**, 2831, 1988.

and bismuth contribute three skeletal electrons; and selenium and tellurium contribute four skeletal electrons, when they occur as bare metal vertices in two- and three-dimensional aromatic systems. Thus Ge, Sn, and Pb vertices are isoelectronic with BH, $Fe(CO)_3$, and C_5H_5Co vertices, and As, Sb, and Bi vertices are isoelectronic with CH, $Co(CO)_3$, and C_5H_5Ni vertices in bare metal cluster compounds. Since bare Ga, In, and Tl vertices contribute only a single skeletal electron, *bare* metal clusters consisting solely of these metals are too electron-poor to be stable; an apparent exception is the recently reported[17] K_8In_{11}, involving different structural ideas discussed later in this chapter.

Some of the structures of the most important bare ionic post-transition metal clusters are depicted in Figure 8–1. Their chemical bonding topologies can be treated as follows:

1. **Square.** Bi_4^{2-}, Se_4^{2+}, and Te_4^{2+} isoelectronic and isolobal with the globally delocalized planar cyclobutadiene dianion with 14 skeletal electrons (e.g., for Bi_4^{2-}: $(4)(3) + 2 = 14$) corresponding to 8 electrons for the 4 σ-bonds and 6 electrons for the π-bonding.
2. **Butterfly.** $Tl_2Te_2^{2-}$ with $(2)(1) + (2)(4) + 2 = 12$ apparent skeletal electrons isoelectronic and isolobal with *neutral* cyclobutadiene but undergoing a Jahn-Teller-like distortion to a butterfly structure as discussed by Burns and Corbett.[18]
3. **Tetrahedron.** $Sn_2Bi_2^{2-}$ and $Pb_2Sb_2^{2-}$ with $(2)(2) + (2)(3) + 2 = 12$ skeletal electrons for localized bonds along the edges of the tetrahedron analogous to organic tetrahedrane derivatives, R_4C_4.
4. **Trigonal Bipyramid.** Sn_5^{2-}, Pb_5^{2-}, and Bi_5^{3+} with 12 skeletal electrons (e.g., $(5)(2) + 2 = 12$ for Sn_5^{2-} and Pb_5^{2-}) analogous to the trigonal bipyramidal $C_2B_3H_5$ carborane.
5. **Seven-vertex Systems.** As_7^{3-} and Sb_7^{3-} with the C_{3v} structure depicted in Figure 8–1. These anions have the correct $(4)(3) + (3)(1) + 3 = 18$ skeletal electrons for edge-localized bonds along the 9 edges derived by considering the three vertices of degree 2 to use two internal orbitals each and the four vertices of degree 3 to use three internal orbitals each, in accord with the requirement for matching the numbers of internal orbitals with the vertex degrees for edge-localized bonding.
6. **Capped Square Antiprism.** Ge_9^{4-}, Sn_9^{4-}, and Pb_9^{4-} with $(9)(2) + 4 = 22 = 2n + 4$ skeletal electrons required for an $n = 9$ vertex C_{4v} *nido* polyhedron having 12 triangular faces and one square face.
7. **Tricapped Trigonal Prism.** Ge_9^{2-} and $TlSn_8^{3-}$ with the $2n + 2 = 20$ skeletal electrons required for an $n = 9$ vertex globally delocalized D_{3h} deltahedron analogous to $B_9H_9^{2-}$ (ref. 19); Bi_9^{5+} anomalously

[17]S. C. Sevov and J. D. Corbett, *Inorg. Chem.*, **30**, 4875, 1991.
[18]R. C. Burns and J. D. Corbett, *J. Am. Chem. Soc.*, **103**, 2627, 1981.
[19]L. J. Guggenberger, *Inorg. Chem.*, **7**, 2260, 1968.

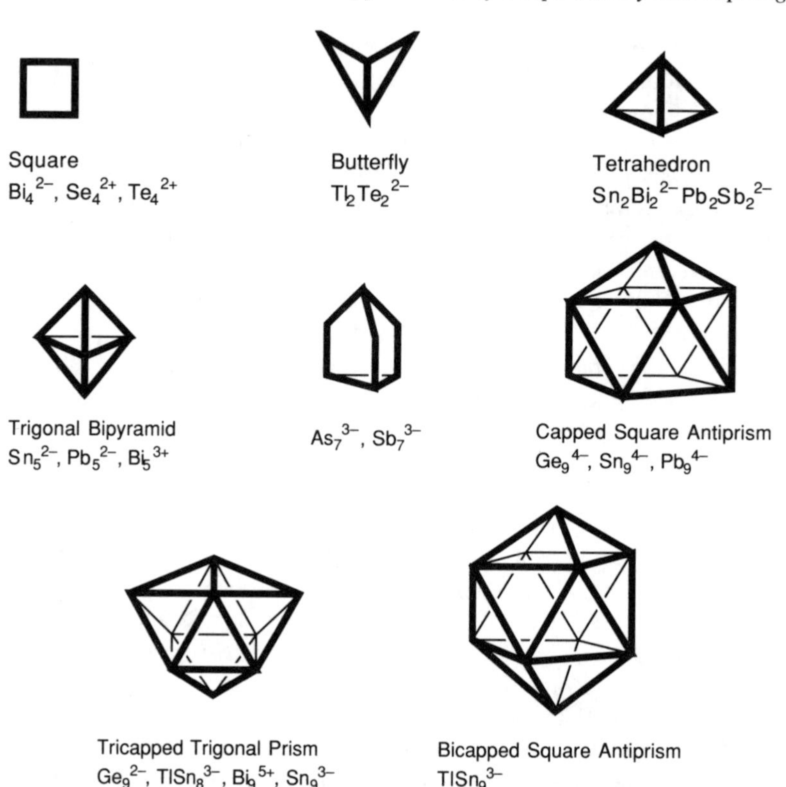

FIGURE 8–1. The shapes of some ionic post-transition metal clusters.

having (9)(3) – 5 = 22 rather than the expected 20 skeletal electrons suggesting[20] incomplete overlap of the unique internal orbitals directed towards the core of the deltahedron; Sn_9^{3-} with (9)(2) + 3 = 21 skeletal electrons including one extra electron for a low-lying antibonding orbital analogous to radical anions formed by stable aromatic hydrocarbons such as naphthalene and anthracene.

8. **Bicapped Square Antiprism.** $TlSn_9^{3-}$ with the (1)(1) + (9)(2) + 3 = 22 = $2n + 2$ skeletal electrons required for n =10 vertex globally delocalized D_{4d} deltahedron (the bicapped square antiprism in Figure 8–1), analogous to that found in the $B_{10}H_{10}^{2-}$ anion.[21]

A completely analogous approach can be used to treat the chemical bonding topology of *gas phase* post-transition metal clusters produced in molecular beam experiments, thereby providing a good demonstration of the

---

[20] R. B. King, *Inorg. Chim. Acta*, **57**, 79, 1982.

[21] R. D. Dobrott and W. N. Lipscomb, *J. Chem. Phys.*, **37**, 1779, 1962.

analogy between condensed phase and gas phase post-transition metal clusters.[22] The bare post-transition metal vertices most suitable as building blocks for structures of gas phase clusters resembling the condensed phase clusters discussed above are Sn and Pb vertices, which are donors of two skeletal electrons each, and Sb and Bi vertices, which are donors of three skeletal electrons each. Experimental information is obtained by the vaporization of samples of these metals or their alloys using resistive heating[7,8,9,10] or laser vaporization.[11,12,13,14,15,16]

1. **Bismuth and Antimony Gas Phase Clusters.** Experiments performed under optimum conditions for the generation of neutral clusters generate predominantly M_2' (M' = Sb, Bi), analogous to the very stable N_2 and M_4' with (4)(3) = 12 skeletal electrons postulated to have tetrahedral structures like P_4. Dominant cationic bismuth clusters include the Bi_3^+ cluster with (3)(3) – 1 = 8 skeletal electrons isoelectronic with the stable cyclopropenyl cation $C_3H_3^+$ and thus postulated to have an equilateral triangle structure; the Bi_5^+ cluster with (5)(3) – 1 = 14 = $2n + 4$ skeletal electrons for $n = 5$ isoelectronic with B_5H_9 and postulated on this basis to have a *nido* square pyramid structure; the Bi_7^+ cluster with the requisite number of skeletal electrons, namely (6)(3) + (1)(5) – 1 = 22 for an edge-localized seven-vertex polyhedron having 11 edges, 6 faces, 6 vertices of degree 3 and 1 vertex of degree 4. Dominant anionic bismuth clusters include Bi_2^- isoelectronic with paramagnetic NO, Bi_3^- isoelectronic with the diamagnetic N_2O, and Bi_5^- isoelectronic with the cyclopentadienide anion $C_5H_5^-$.
2. **Tin and Lead Gas Phase Clusters.** Since tin and lead vertices using the normal three internal orbitals each are sources of two skeletal electrons, all neutral homoatomic tin and lead clusters with n vertices necessarily have $2n$ skeletal electrons and thus are electron-poor systems expected to be capped deltahedra (Section 6.3). Molecular beam experiments on the generation of such clusters, using either resistive heating or laser vaporization, generate local maxima at M_7 and M_{10} (M = Sn, Pb), corresponding to the capped octahedron and 3,4,4,4-tetracapped trigonal prism structures. Each of these two structures has C_{3v} symmetry and one tetrahedral chamber with an octahedron and a 4,4,4-tricapped trigonal prism, respectively, as the central deltahedron. The apparent preference of these electron-poor metal clusters to form capped deltahedra having threefold symmetry may account for these observed abundances.
3. **Mixed Sn/Bi, Pb/Sb, and Sn/As Clusters.** The nine-atom clusters, Sn_5Bi_4, $Sn_4Bi_5^+$, Pb_5Sb_4, $Pb_4Sb_5^+$, Sn_5As_4, and $Sn_4As_5^+$,

---

[22] R. B. King, *J. Phys. Chem.*, **92**, 4452, 1988.

each with 22 skeletal electrons isoelectronic with the condensed phase clusters Ge_9^{4-}, Sn_9^{4-}, and Pb_9^{4-}, shown to have capped square antiprism structures (Figure 8–1), dominate in these mixed metal systems. *These mixed-metal systems represent a critical link between gas-phase and condensed-phase post-transition metal clusters.* The Pb/Sb and Sn/As systems also exhibit definite local maxima at the $2n + 2$ skeletal electron species $M_3M'_2$ and $M_4M'_2$ (M = Sn, Pb; M' = As, Sb) isoelectronic with the carboranes $C_2B_3H_5$ and $C_2B_4H_6$.

4. **Mixed In/Sb and In/Bi Clusters.** Dominant cations in the In/Sb system are $InBi_4^+$ with $(1)(1) + (4)(3) - 1 = 12$ skeletal electrons isoelectronic with the trigonal bipyramid condensed phase Sn_5^{2-}, Pb_5^{2-}, and Bi_5^{3+} ions and $In_2Bi_7^+$ with $(2)(1) + (7)(3) - 1 = 22$ skeletal electrons isoelectronic with the capped square antiprism condensed phase species Ge_9^{4-}, Sn_9^{4-}, and Pb_9^{4-}, as well as with gas phase species such as Sn_5Bi_4 and $Sn_4Bi_5^+$. The mixed In/Sb and In/Bi clusters thus provide additional links between gas-phase and condensed-phase metal clusters. In addition, a dominant neutral species in the In/Sb system is neutral In_2Sb_4 with $(2)(1) + (4)(3) = 14$ skeletal electrons isoelectronic with the neutral gas phase cluster Pb_4Sb_2, as well as the boranes $C_2B_4H_6$ and $B_6H_6^{2-}$, shown to have octahedral structures.

8.2 POLYHEDRAL GALLIUM AND INDIUM CLUSTERS

The elements gallium, indium, and thallium have 13 valence electrons. Bare gallium, indium, and thallium vertices are therefore donors of only a single skeletal electron. For this reason, clusters composed exclusively of bare gallium, indium, and thallium vertices are usually too electron-poor to be stable, even when additional electrons are added by reduction with alkali metals to form anionic species. The sole exception to this rule is the recently discovered[17] indium cluster K_8In_{11}, which contains a C_{3v} deltahedron of 11 bare indium atoms (Figure 8–2). Gallium clusters are found in gallium-rich intermetallic phases of gallium and alkali metals. Belin and Ling[23] have shown that the electron counting methods discussed in earlier chapters for the study of boron polyhedra (Chapter 4) and transition metal clusters (Chapter 6) can also be applied to these gallium intermetallic phases. Burdett and Canadell[24,25] have also discussed structural chemistry and electron counting in gallium clusters.

The gallium vertices in the gallium-rich intermetallic phases of gallium and alkali metals, like other post-transition metal vertices, may be regarded as

---

[23] C. Belin and R. G. Ling, *J. Solid State Chem.*, **48**, 40, 1983.
[24] J. K. Burdett and E. Canadell, *J. Am. Chem. Soc.*, **112**, 7207, 1990.
[25] J. K. Burdett and E. Canadell, *Inorg. Chem.*, **30**, 1991, 1991.

having an sp^3d^5 manifold of nine valence orbitals, but with all five of the *d* orbitals having non-bonding electron pairs. Thus, only the four orbitals from the sp^3 manifold participate in the metal cluster bonding, as is the case for boron vertices (Chapter 4). Among these four valence orbitals, three are normally used for intrapolyhedral skeletal bonding, leaving the fourth orbital for bonding to an external group, usually a two-center bond to a gallium vertex of an adjacent deltahedron or to an extra "satellite" gallium atom between the gallium deltahedra. These satellite gallium atoms can be joined in pairs (e.g., in $RbGa_7$ and Li_3Ga_{14}) or even into complicated open polyhedra (e.g., a Ga_{15} polyhedron (Figure 8–2) in $Na_{22}Ga_{39}$). Exceptionally, the external orbitals of the gallium vertices of the deltahedra can form three-center Ga_3 bonds (e.g., in $RbGa_7$) or can contain a non-bonding lone electron pair (e.g., three of the gallium vertices in one of the types of gallium icosahedra in $Na_{22}Ga_{39}$), like all of the vertices in the post-transition metal clusters discussed in Section 8.1. A gallium vertex contributing one electron to a two-center external bond functions as a donor of two skeletal electrons, like the BH vertices in $B_nH_n^{2-}$ deltahedra (Chapter 4). A gallium vertex with a lone pair in the external orbital functions as a donor of only a single skeletal electron, consistent with the rules for post-transition metal clusters discussed above.

The structures of the intermetallic phases of gallium and alkali metals consist of infinite networks of linked gallium polyhedra with alkali metals in some of the interstices. The alkali metals are assumed to form monopositive ions, thereby donating one electron to the gallium network for each alkali metal atom. This pure ionic model breaks down somewhat for the lithium/gallium phases, which may have fewer electrons for the pure closed shell configurations. This may be attributed to the tendency for lithium to form partially covalent $LiGa_n$ multicenter bonds, so that the valence electron from a single lithium atom is effectively shared with more than one gallium atom.

A three-dimensional structure consisting of linked gallium deltahedra and alkali-metal atoms still has holes in it that can be filled by additional gallium atoms which are *not* vertices of the deltahedra. Such "satellite" gallium atoms typically form four two-electron two-center bonds and thus bear a formal negative charge. Satellite gallium atoms can also be bonded to other satellitegallium atoms as well as to gallium atoms that are vertices of deltahedra. In $Na_{22}Ga_{39}$, the 15-vertex polyhedron formed by the satellite gallium atoms (Figure 8–2), which has mainly degree 3 vertices, may be regarded as an edge-localized polyhedron in contrast to the globally delocalized deltahedra.

Using this general approach, the following intermetallic phases of gallium and alkali metals can be considered:

Bisdisphenoid
("D_{2d} Dodecahedron")
Ga_8

Edge-coalesced Icosahedron
Ga_{11}

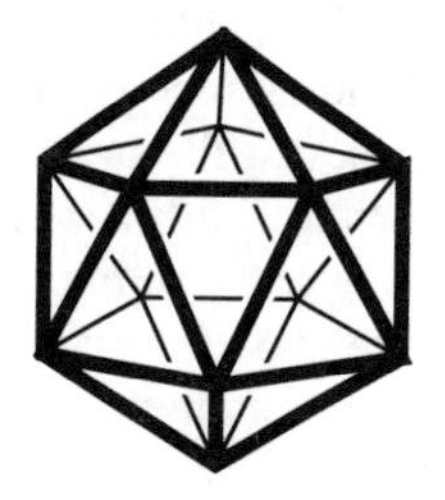

Regular Icosahedron
Ga_{12}

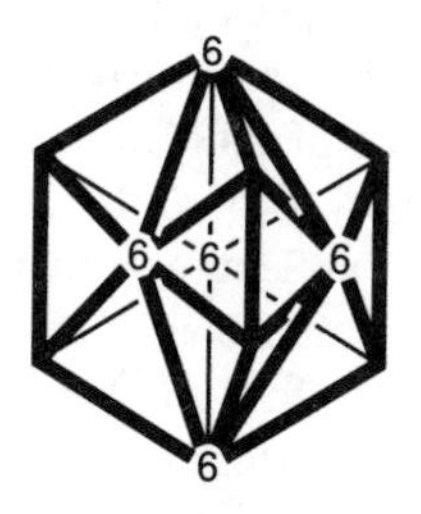

11-vertex Deltahedron
In_{11}

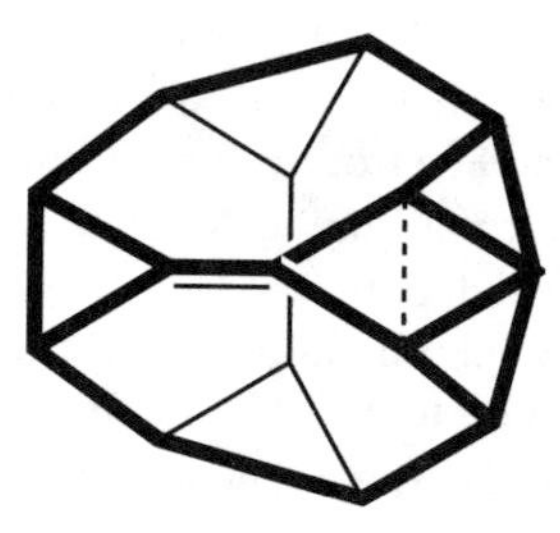

Ga_{15} satellite polyhedron
found in $Na_{22}Ga_{39}$

FIGURE 8–2. Gallium and indium polyhedra found in intermetallic phases with alkali metals. The vertices of degree 6 are indicated on the two 11-vertex deltahedra.

1. **KGa_3, $RbGa_3$.** These phases[26] may be represented as $M_3(Ga_8)(Ga)$, in which the Ga_8 deltahedron is a bisdisphenoid (Figure 8–2) with four degree 4 vertices and four degree 5 vertices. The degree 5 gallium vertices are linked to the degree 5 gallium vertices of adjacent bisdisphenoids (Ga–Ga = 2.61–2.74 Å), whereas the degree 4 vertices are linked to the degree 4 vertices of adjacent bisdisphenoids through four-coordinate satellite gallium atoms. The closed-shell electronic

[26] R. G. Ling and C. Belin, *Z. anorg. allgem. Chem.*, **480**, 181, 1981.

configuration for Ga_8^{2-} for the bisdisphenoids with two-center external bonds from each vertex atom and Ga^- for the four-coordinate satellite gallium atoms leads to the observed stoichiometry $M_3(Ga_8)(Ga) = MGa_3$.

2. **$RbGa_7$, $CsGa_7$.** These phases[27] may be represented as $M_2(Ga_{12})(Ga_2)$, with a Ga_{12} icosahedron and a bonded pair (Ga–Ga = 2.52 Å) of four-coordinate satellite gallium atoms. Six of the gallium vertices of each Ga_{12} icosahedron are bonded externally to other gallium atoms through two-center two-electron bonds, whereas the other six gallium vertices are bonded externally to other gallium atoms through three-center two-electron bonds. In such a situation, the $2n + 2 = 26$ skeletal electron closed-shell electronic configuration of a gallium icosahedron is not the usual Ga_{12}^{2-} but instead *neutral* Ga_{12}, determined as follows:

Valence electrons of 12 Ga atoms: (12)(3) =	36 electrons
Required for 6/2 external two-center two-electron bonds: (6/2)(2) =	–6 electrons
Required for 6/3 external three-center two-electron bonds: (6/3)(2) =	–4 electrons
Net electrons remaining for skeletal electrons:	*26 electrons*

The closed-shell electronic configurations of Ga_{12} for the gallium icosahedra and Ga_2^{2-} for the bonded pairs of four-coordinate satellite atoms lead to the observed stoichiometry $M_2(Ga_{12})(Ga_2) = MGa_7$.

3. **Li_3Ga_{14}.** This phase[28] may be represented as $Li_3(Ga_{12})(Ga_2)$ with a Ga_{12} icosahedron and a bonded pair (Ga–Ga = 2.61 Å) of four-coordinate satellite gallium atoms. In each gallium icosahedron, six of the vertices are linked directly to vertices of adjacent gallium icosahedra, and the other six vertices are linked to the satellite gallium atoms in all cases by two-center two-electron bonds leading to the Ga_{12}^{2-} closed shell electronic configuration. The overall Li_3Ga_{14} stoichiometry is one electron short per formula unit of the closed shell electronic configuration $(Ga_{12}^{2-})(Ga_2^{2-}) = Ga_{14}^{4-}$, assuming complete ionization of the lithium to Li^+. However, as noted above, multicenter $LiGa_n$ bonding may relieve such apparent electron deficiencies.
4. **K_3Ga_{13}.** This phase[29] may be represented as $K_6(Ga_{11})(Ga_{12})(Ga)_3$, with equal quantities of Ga_{11} and Ga_{12} deltahedra (Figure 8–2) and three satellite gallium atoms for each 23 (i.e., $Ga_{11} + Ga_{12}$) deltahedral gallium atoms. Each vertex atom in both types of gallium deltahedra forms one two-center two-electron external bond leading to the closed

[27] C. Belin, *Acta Cryst.*, **B37**, 2060, 1981.
[28] C. Belin and R. G. Ling, *J. Solid State Chem.*, **45**, 290. 1982.
[29] C. Belin, *Acta Cryst.*, **B36**, 1339, 1980.

shell electronic configurations Ga_{11}^{2-} and Ga_{12}^{2-}. There are both three-coordinate and four-coordinate satellite gallium atoms, with twice as many four-coordinate satellite gallium atoms as three-coordinate satellite gallium atoms. The electronic configurations of Ga_{11}^{2-} and Ga_{12}^{2-} for the gallium deltahedra, Ga^{-} for the four-coordinate satellite gallium atoms, and Ga for the three-coordinate satellite gallium atoms lead to the observed stoichiometry $K_6(Ga_{11})(Ga_{12})(Ga^{3\text{-coord}})(Ga^{4\text{-coord}})_2 = K_6Ga_{26} = K_3Ga_{13}$.

5. **$Na_{22}Ga_{39}$.** This phase[30] may be represented as $Na_{22}(Ga_{12})_2(Ga_{15})$, with two different types of Ga_{12} icosahedra in equal quantities and open Ga_{15} polyhedra of satellite gallium atoms with the topology depicted in Figure 8–2. Except for the relatively long Ga–Ga edge (2.99 Å), depicted by a dotted line in Figure 8–2, which may not represent a gallium-gallium bond, the Ga_{15} polyhedron of satellite gallium atoms has 14 vertices of degree 3 and only one vertex of degree 4, suggestive of edge-localized bonding using three of the four valence orbitals of the sp^3 manifolds from each gallium vertex. In addition, all but two of the gallium vertices of the Ga_{15} satellite polyhedron are bonded through two-center two-electron bonds to exactly one gallium vertex of another polyhedron. The two gallium vertices not bonded to gallium atoms of another polyhedron form an unusually short bond (Ga–Ga = 2.43 Å), which suggests a Ga=Ga double bond (depicted in Figure 8–2). Using this interpretation, the closed-shell electronic configuration for the edge-localized gallium satellite polyhedron depicted in Figure 8–2 is Ga_{15}^{15-}.

There are two types of Ga_{12} icosahedra in $Na_{22}Ga_{39}$. In one type of Ga_{12} icosahedron (icosahedra A), all 12 vertices form two-center two-electron external bonds to gallium vertices either of other icosahedra or of the satellite Ga_{15} polyhedron leading to the usual closed shell electronic configuration Ga_{12}^{2-}. However, in the other type of Ga_{12} icosahedron in $Na_{22}Ga_{39}$ (icosahedra B), three of the 12 vertices are not bonded to external groups, leading to the 26 skeletal electron Ga_{12}^{5-} closed-shell electronic configuration as follows:

Valence electrons of 12 Ga atoms: (12)(3) =	36 electrons
–5 charge on Ga_{12}^{5-}:	5 electrons
Required for 9/2 external two-center two-electron bonds: (9/2)(2) =	–9 electrons
Required for 3 external lone pairs:	–6 electrons
Net electrons remaining for skeletal electrons:	*26 electrons*

The closed-shell electronic configurations of Ga_{12}^{2-} for icosahedra A, Ga_{12}^{5-}

[30]R. G. Ling and C. Belin, *Acta Cryst.*, **B38**, 1101, 1982.

for icosahedra B, and Ga_{15}^{15-} for the edge-localized gallium satellite polyhedron (Figure 8–2) lead to the observed stoichiometry $Na_{22}(Ga_{12})_2(Ga_{15})$ = $Na_{22}Ga_{39}$.

Much less information is available on indium clusters in intermetallic phases of indium and alkali metals. Thus, only one cluster containing an indium polyhedron has been structurally characterized, namely K_8In_{11} (ref. 17). The structure of K_8In_{11} is based on an 11-vertex deltahedron of *bare* indium atoms, which therefore must be regarded as donors of one skeletal electron each, if considered as normal indium vertices using three internal orbitals each. The topology of the 11-vertex deltahedron in K_8In_{11} is very different from that of the 11-vertex deltahedron known as the edge-coalesced deltahedron, found not only in K_3Ga_{13} discussed above but also in the deltahedral borane anion $B_{11}H_{11}^{2-}$ and the corresponding carborane $C_2B_9H_{11}$.[31] Thus, although both 11-vertex polyhedra are deltahedra having 18 faces, all of which are triangles, the edge-coalesced icosahedron in $B_{11}H_{11}^{2-}$ and $C_2B_9H_{11}$ has only one vertex of degree 6 as well as eight vertices of degree 5 and two vertices of degree 4, whereas the 11-vertex deltahedron in K_8In_{11} has five vertices of degree 6 and six vertices of degree 4 (Figure 8–2). Neither of these 11-vertex deltahedra have any degree 3 vertices, and the edge-coalesced icosahedron minimizes the number of degree 6 vertices.[32]

The skeletal electron count of K_8In_{11} poses an interesting question. If its 11-vertex deltahedron is considered to be a globally delocalized polyhedron, the 11 normal indium vertices contribute a total of 11 skeletal electrons and the –8 charge contributes another 8 skeletal electrons, for a total of 19 skeletal electrons, which is five electrons short of the $2n + 2 = 24$ skeletal electrons for $n = 11$, thereby indicating an electron-poor system. The lack of degree 3 vertices means that the 11-vertex deltahedron in K_8In_{11} does not have any of the tetrahedral chambers found in electron-poor transition metal carbonyl clusters (Section 6.3). However, the 11-vertex deltahedron in K_8In_{11} can be dissected into a central In_7 pentagonal bipyramid with two bonded pairs of capping vertices. Such a bonding topology with a globally delocalized central In_7 pentagonal bipyramid requires a total of 20 skeletal electrons distributed as follows:

Surface bonding in the central In_7 pentagonal bipyramid: (2)(7) =	14 electrons
7-center core bonding in the central In_7 pentagonal bipyramid:	2 electrons
2 two-electron two-center bonds involving pairs of capping In atoms:	4 electrons
Total skeletal electrons required:	*20 electrons*

[31]F. Klanberg and E. L. Muetterties, *Inorg. Chem.*, **5**, 1955, 1966.

[32]R. B. King and A. J. W. Duijvestijn, *Inorg. Chim. Acta*, **178**, 55, 1990.

The actual observed 19 skeletal electrons is thus only one electron short of the closed shell electronic configuration. This electron deficiency can account for the observed[17] metallic behavior of this cluster.

8.3 MERCURY VERTICES IN METAL CLUSTERS: ALKALI METAL AMALGAMS

Mercury has 12 valence electrons. If mercury uses a nine-orbital sp^3d^5 manifold and the normal three internal orbitals, then all 12 mercury valence electrons are used to fill the six external orbitals, so that a bare mercury vertex is a donor of zero skeletal electrons. Clusters constructed solely of neutral mercury atoms are therefore not stable because of the absence of bonding electrons. An obvious consequence of this is the monoatomic nature of elemental mercury, thereby making it by far the most volatile of all metals. However, reduction of mercury to an anionic species provides a source of bonding electrons, so anionic mercury clusters can be stable. Anionic mercury clusters are therefore found in alkali metal amalgams in which all of the bonding electrons are effectively furnished by the alkali metal.

Several alkali metal amalgams have been structurally characterized, including the sodium amalgams[33] $NaHg_2$, NaHg, and Na_3Hg_2 and the potassium amalgams[34] KHg_2 and KHg. The most highly reduced of these species, namely Na_3Hg_2, has been shown[35] to contain discrete 2.99 Å square planar Hg_4^{6-} clusters which are *not* isoelectronic with other square planar post-transition metal clusters such as Bi_4^{2-}, Se_4^{2+}, and Te_4^{2+}. In gold-colored NaHg, slightly distorted 3.05 × 3.22 Å Hg_4 rectangles are fused into a zigzag ribbon, whereas in the likewise gold colored KHg, slightly distorted (93.5° rather than 90°) 3.03 Å Hg_4 squares are linked by 3.36 Å Hg–Hg bonds. The structure of $Rh_{15}Hg_{16}$ has Hg_8 cubes (Hg–Hg edges in the range 2.95 to 3.04 Å).[36] More complicated networks of mercury rectangles and parallelograms are present in $NaHg_2$ and KHg_2. Thus, the prototypical building blocks for alkali metal amalgams are derived from the Hg_4^{6-} squares found in Na_3Hg_2. Electrochemical studies[37,38] suggest that similar mercury cluster anions are present in "quaternary ammonium amalgams", although difficulties in obtaining crystals from these air- and moisture-sensitive systems have precluded their definitive structural characterization.

The mercury atoms in anionic mercury clusters, such as those found in

[33]J. W. Nielsen and N. C. Baenziger, *Acta. Cryst.*, **7**, 277, 1954.

[34]E. J. Duwell and N. C. Baenziger, *Acta Cryst.*, **8** 705.,1955)

[35]J. D. Corbett, *Inorg. Nucl. Chem. Lett.*, **5**, 81, 1969.

[36]H.-J. Deiseroth and A. Strunck, *Angew. Chem. Int. Ed.*, **28**, 1251, 1989.

[37]E. Garcia, A. H. Cowley, and A. J. Bard, *J. Am. Chem. Soc.*, **108**, 6082, 1986.

[38]E. Kariv-Miller and V. Svetličić, *J. Electroanal. Chem.*, **205**, 319, 1986.

alkali metal amalgams, can be considered to have seven-orbital spd^5 cylindrical bonding manifolds, similar to the vertex gold atoms in the peripheral polyhedra of the centered gold clusters (Section 7.2). Six of these seven orbitals, namely the *s* orbital and the five *d* orbitals, are external orbitals, whereas the single *p* orbital is an internal orbital. The chemical bonding topology of the mercury atoms in anionic mercury clusters can be related to the spectrum of the graph describing the topology of the overlap of these internal orbitals using the method discussed in Section 4.1. Thus, the mercury square found in Na_3Hg_2 has one positive eigenvalue and two zero eigenvalues whereas the cube found in $Rb_{15}Hg_{16}$ has four positive eigenvalues (Figure 8–3).

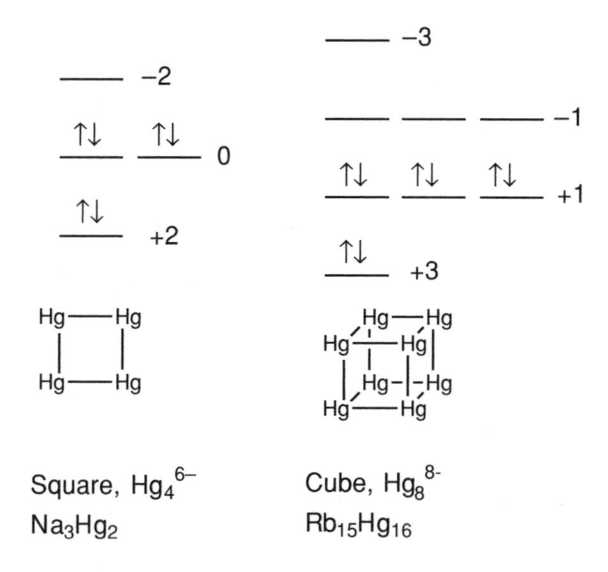

FIGURE 8–3. The spectra of the square found in the Hg_4^{6-} anion in Na_3Hg_2 and the cube found in the Hg_8^{8-} anion in $Rb_{15}Hg_{16}$.

Consider first the mercury square found in the Hg_4^{6-} anion of Na_3Hg_2. Since a neutral mercury atom with six external orbitals for its 12 valence electrons is a donor of zero skeletal electrons, the Hg_4^{6-} anion has only the six skeletal electrons arising from the –6 charge. These six skeletal electrons are exactly enough electrons to fill the single bonding and two non-bonding orbitals of the mercury square, as indicated in Figure 8–3. Similarly, each mercury cube in the "mercubane" cluster $Rb_{15}Hg_{16}$ has 7.5 skeletal electrons, which is only 0.5 skeletal electron short of the eight skeletal electrons required to fill all four of its bonding orbitals (Figure 8–3).

Chapter 9

INFINITE SOLID STATE STRUCTURES WITH METAL-METAL INTERACTIONS

9.1 FUSION OF METAL CLUSTER OCTAHEDRA INTO BULK METAL STRUCTURES

Section 7.1 discusses the chemical bonding topology of octahedral early transition metal halide clusters. Such early transition metal octahedra can be used as building blocks for infinite solid state structures by infinite fusion, first in one dimension to give chains (e.g., Gd_2Cl_3) and then in two dimensions to give sheets (e.g., ZrX). The limiting case of infinite fusion in all three dimensions corresponds to bulk metals. The halogen/metal ratio in such infinite solid state structures decreases as the number of infinite dimensions (i.e., the "dimensionality") increases (e.g., 2 in the "Mo_6Cl_{12}" discrete octahedron, 1.5 in the Gd_2Cl_3 infinite chain structure, 1 in ZrX, and 0 in bulk metals. The fusion of such early transition metal halide octahedral cluster building blocks involves features different from the fusion of octahedral metal clusters carbonyls discussed in Section 6.6.

The skeletal chemical bonding topology of a metal cluster, such as those discussed in Chapters 6 to 8, consists of a set containing two-electron two-center bonds and/or two-electron multicenter bonds; the latter involve overlap of more than two valence orbitals for each bond. Satisfactory electron counting schemes for a given skeletal bonding topology require allocation of the available skeletal electrons and internal orbitals to the individual bonds, normally using two electrons for each bond and n orbitals for an n-center bond. Difficulties can arise from uncertainties in the valence orbital manifolds and hence in the electronic configurations of the vertex atoms or in the partition of vertex atom orbitals between internal orbitals participating in the skeletal bonding and external orbitals participating in bonding external to the cluster. Such difficulties can sometimes lead to ambiguities in the assignments of bonding topologies in metal clusters in cases where two or more different skeletal bonding topologies assign reasonable electron configurations to the vertex atoms and use all of the available orbitals and electrons. Such ambiguities arise relatively rarely in the treatment of discrete molecular and ionic metal clusters (such as those discussed in Chapters 6 to 8) but occur more frequently in the infinite solid state structures discussed in this chapter. This relates to the coalescence of the discrete energy levels of distinct molecular orbitals in finite structures into broad bands in infinite solid state structures. In addition, in order to provide partially filled valence bands, highly conducting infinite metal clusters may not have enough skeletal electrons to fill all of the bonding orbitals. These ambiguities limit the applicability of graph-theory derived methods for the study of the bonding topologies of infinite metal clusters such as those discussed in this chapter, with the difficulties apparently increasing as the number of dimensions of infinite delocalization increases. Nevertheless, ideas leading to satisfactory chemical bonding models for discrete metal clusters still lead to consistent results for infinite metal clusters. Other methods for the treatment of infinite solid state metal clusters based on topological ideas include the method of moments developed by Burdett and co-workers.[1,2,3] This method will not be discussed in this book.

Many infinite solid state structures constructed from metal-metal bonds are based on an octahedral metal cluster repeating unit. Atoms shared by two or more such repeating units are partitioned equally between the repeating units. Electrons and orbitals from such shared atoms may not necessarily be partitioned equally, but the sums of the electrons and the orbitals donated by the atom to all of the units sharing the atom in question must equal the number of valence electrons and orbitals available from the neutral atom, generally three or four electrons and eight or nine orbitals.

One-dimensional infinite chains of edge-fused octahedra occur in the lanthanide halides of stoichiometry M_2Cl_3, as exemplified by Gd_2Cl_3 (Figure 9–1).[4] The metal chains in this structure have both Gd_6 ($Gd_2{}^aGd_4{}^b$) octahedral cavities and Gd_4 ($Gd_2{}^aGd_2{}^b$) tetrahedral cavities, with twice the number of tetrahedral cavities as octahedral cavities. A repeating octahedral Gd_6 unit in the chain can be represented as $Gd_2{}^aGd_{4/2}{}^b(\mu_3{}^{abb}\text{-}Cl)_4(\mu_2{}^{aa}\text{-}Cl)_{4/2}$, in which a subscript "$_{4/2}$" refers to four atoms each shared with two metal octahedra. The coordination polyhedron of the bridging gadolinium atoms (Gd^b in Figure 9–1) may be approximated by a capped square antiprism (Figure 7–1), having an external halogen atom in the axial position, bonds to $\mu_3{}^{abb}$ face-bridging halogen atoms in the four medial positions, and internal orbitals in the four basal positions. The axial gadolinium atoms (Gd^a in Figure 9–1) are also nine-coordinate, having four internal orbitals, a bond to an external halogen atom, two bonds to halogen atoms bridging an aa-edge of the tetrahedral cavity to an adjacent octahedron, and two bonds to $\mu_3{}^{abb}$ face-bridging halogen atoms in the same octahedron. Both the axial and bridging gadolinium atoms thus have five external orbitals and are $(2)(5) - 3 - 2 = 5$ electron acceptors (i.e., –5 electron donors), after allowing for the three electrons of the neutral gadolinium and an electron pair from the external atoms (not shown for clarity in Figure 9–1). This leads to the following count of skeletal electrons and internal orbitals for an octahedral Gd_6 unit in Gd_2Cl_3 ($= Gd_2{}^aGd_{4/2}{}^b(\mu_3{}^{abb}\text{-}Cl)_4(\mu_2{}^{aa}\text{-}Cl)_{4/2}$:

[1] J. K. Burdett, S. Lee, and W. C. Sha, *Croat. Chim. Acta*, **57**, 1193, 1984.
[2] J. K. Burdett and S. Lee, *J. Solid State Chem.*, **56**, 211, 1985.
[3] J. K. Burdett, *Acc. Chem. Res.*, **21**, 189, 1988.
[4] D. A. Lokken and J. D. Corbett, *Inorg. Chem.*, **12**, 556, 1973.

2 axial Gd (Gd^a): (2)(–5) =	–10 electrons	8 orbitals
4/2 bridging Gd (Gd^b): (4/2)(–5) =	–10 electrons	8 orbitals
4 μ_3^{abb}-Cl: (4)(5) =	20 electrons	
4/2 μ_2^{aa}-Cl: (4/2)(3) =	6 electrons	
Total skeletal electrons and internal orbitals:	*6 electrons*	*16 orbitals*

These skeletal electrons and internal orbitals can be used in the octahedral Gd_6 unit as follows:

1 six-center octahedral core bond (a^2b^4)	2 electrons	6 orbitals
2 four-center tetrahedral core bonds (a^2b^2)	4 electrons	8 orbitals
Total skeletal electrons and orbitals required:	*6 electrons*	*14 orbitals*

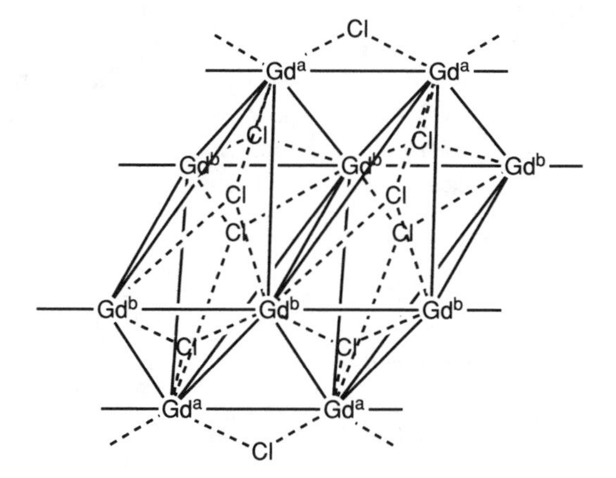

FIGURE 9–1. The building block in infinite chain lanthanide halide structures (e.g., Gd_2Cl_3).

The failure to use two of the available 16 orbitals in this bonding topology corresponds to the axial gadolinium atoms (two for each octahedral Gd_6 unit) having 16-electron rather than 18-electron configurations. Also the tendency for core bonding in tetrahedral as well as octahedral cavities in even this one-dimensional infinite metal cluster contrasts with the edge-localized bonding always found in tetrahedral chambers in discrete metal clusters (see Section 6.3). The closed shell electronic configuration of Gd_2Cl_3 obtained by the electron-counting scheme outlined above is consistent with its semiconducting energy gap E_g of approximately 1 eV.[5]

Two-dimensional infinite sheets of metal octahedra occur in the graphite-like zirconium monohalides.[6] The structures of these systems are built from two layers of hexagonal sheets of metal atoms which form both octahedral

[5]D. W. Bullett, *Inorg. Chem.*, **24**, 3319, 1985.

[6]D. G. Adolphson and J. D. Corbett, *Inorg. Chem.*, **15**, 1820, 1976.

and tetrahedral cavities. There are twice as many tetrahedral as octahedral cavities in these infinite sheet structures as in the infinite chain structure of Gd_2Cl_3 discussed above. The octahedral cavities each have six internal faces and two external faces; the external faces are capped by μ_3 face-bridging halogen atoms (Figure 9–2). Each metal vertex is shared by three octahedral cavities. The lattice symmetry requires the coordination polyhedra of the metal vertices to have three-fold symmetry, and therefore the 4,4,4-tricapped trigonal prism with D_{3h} symmetry is used (Figure 4–5). The nine valence orbitals of each vertex metal atom are partitioned into three groups of three orbitals each, again reflecting the three-fold symmetry of the metal coordination. Thus, for each vertex metal atom, three external orbitals are used for bonds with face-bridging halogen atoms, three internal orbitals are used for core bonding in the three octahedral cavities meeting at the metal vertex in question, and three internal orbitals are used for face bonding across the external faces of the tetrahedral cavities meeting at the metal vertex in question. A zirconium vertex in such a structure uses three external orbitals and is an acceptor of (2)(3) – 4 = 2 skeletal electrons (i.e., a –2 skeletal electron donor) after allowing for the four valence electrons of the neutral zirconium atom. This leads to the following electron counting scheme for ZrCl (= $Zr_{6/3}(\mu_3\text{-}Cl)_2$):

6/3 Zr: (6/3)(–2) =	–4 electrons	12 orbitals
2 μ_3-Cl: (2)(5) =	10 electrons	
Total skeletal electrons and orbitals:	*6 electrons*	*12 orbitals*

These six skeletal electrons and 12 internal orbitals can be used for the following bonding topology based on a single repeating $Zr_{6/3}(\mu_3\text{-}Cl)_2$ repeating unit:

1 six-center octahedral core bond:	2 electrons	6 orbitals
2 three-center face bonds across external faces of tetrahedral cavities:	4 electrons	6 orbitals
Total skeletal electrons and orbitals:	*6 electrons*	*12 orbitals*

The hydrogen-stabilized lanthanide monohalides[7,8] of the stoichiometry HLnX are isoelectronic and isostructural with the zirconium monohalides and can therefore have similar chemical bonding topologies with analogous electron and orbital counts.

The methods used to treat the chemical bonding topology in the infinite chains of Gd_2Cl_3 and the infinite sheets of ZrCl can be extended to the treatment of bulk metals, especially early transition metals, as infinite arrays of fused octahedra in all three dimensions. The structures can be visualized as

---

[7]H. Mattausch, A. Simon, N. Holzer, and R. Eger, *Z. anorg. allgem. Chem.*, **466**, 7, 1980.
[8]A. Simon, *J. Solid State Chem.*, **57**, 2, 1985.

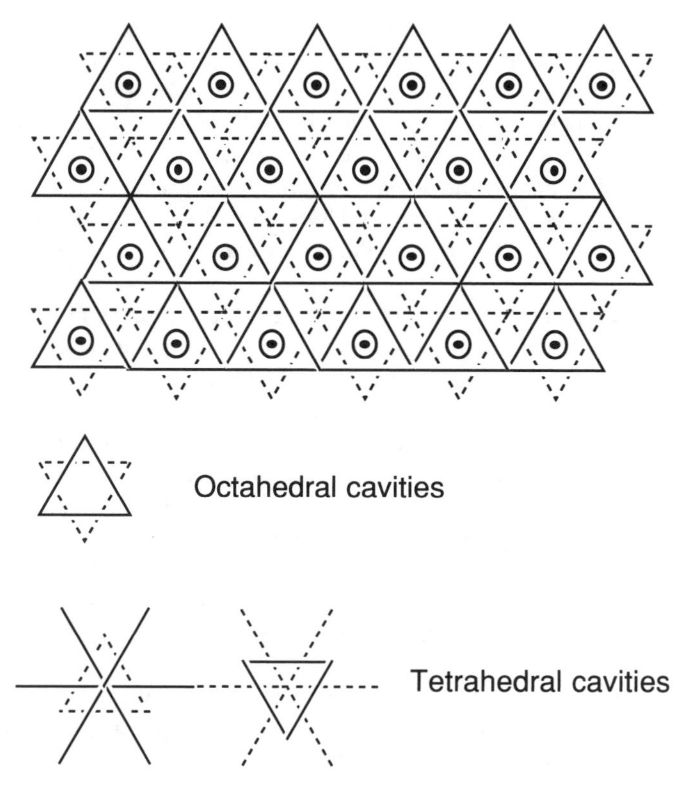

FIGURE 9–2. A top view of a segment of the two stacked hexagonal sheets of metal atoms in the zirconium monohalide structure. The sheet indicated in dotted lines is below the sheet indicated in solid lines. Circled dots indicate the sites of face-bridging halogen atoms above and below the sheets.

an infinite stacking of the hexagonal metal sheets in Figure 9–2 into the third dimension perpendicular to the sheets. In frequently encountered metallic structures, such as the cubic close packed structures, there are two tetrahedral cavities for each octahedral cavity, as in the infinite one-dimensional edge-fused chains of metal octahedra (e.g., Gd_2Cl_3 in Figure 9–1) and infinite edge-fused sheets of metal octahedra (e.g., ZrCl in Figure 9–2) discussed above. In a bulk metal, all of the metal valence orbitals are internal orbitals. Since each metal atom is shared by six octahedral cavities and since an octahedral cavity is formed by six metal atoms, the number of valence electrons for each octahedral cavity is equal to the number of valence electrons of the metal. Formation of one multicenter bond each in each octahedral cavity and in the two tetrahedral cavities for each octahedral cavity requires six electrons per octahedral cavity, corresponding to a metal atom with six valence electrons such as chromium, molybdenum, and tungsten.

This method of treating the chemical bonding topology of bulk metals indicates a special role of the group 6 metals chromium, molybdenum, and tungsten. Such a special role of these metals as forming a "transition metal divide" was recognized by the metallurgist H. E. N. Stone[9] in a study using the "long form" of the periodic table of the systematics of a number of metallic and alloy phases, including iron alloys, β-tungsten phases, lanthanide alloys, and Laves phases such as UCo_2. Stone's approach places the following four "divides" in the periodic table as indicated in Figure 9–3:

1. The "ionic divide" is familiar to inorganic chemists as the rare gases which separate the elements forming anions (e.g., the halogens) from the elements forming cations (e.g., the alkali metals). The rare gases represent a minimum of chemical reactivity.
2. The "covalent divide" is located in the C, Si, Ge, Sn, Pb column which separates the compositions of p-type semiconductors with holes as carriers from n-type semiconductors with electrons as carriers. The electronic configuration of the elements in the "covalent divide" represents a minimum in conductivity.
3. The "composite divide" is located at the coinage metals and can be related to the Hume-Rothery approach for the structure of metals and alloys such as brasses.[10]
4. The "transition metal divide" is located in the Cr, Mo, W column. The electronic configurations of these metals represent a minimum superconducting T_c (see next section).

These divides are placed in such positions that any element near to and on one side of a divide will form a compound or compounds with a counterpart element on the opposite side of that divide; if two elements are near to but on the same side of a divide, only less stable compounds or no compounds at all are formed. In the latter case involving only metallic elements, no intermetallic compounds are formed, and the system has some other form of phase diagram ranging from solid solubility to liquid insolubility.

In his treatment of the transition metal and composite divides (the divides least familiar to conventional inorganic chemists) Stone[9] makes the following two points:

1. The "zone of influence" of a divide does not generally coincide with the block of elements up to the adjacent divides on either side. Sometimes it extends further than an adjacent divide, sometimes less, but nearly always it extends two elements to each side of a divide.

---

[9] H. E. N. Stone, *Acta Metallurgica*, **27**, 259, 1979.

[10] W. Hume-Rothery, R. E. Smallman, and C. W. Haworth, *The Structure of Metals and Alloys*, The Metals and Metallurgy Trust, London, 1969.

2. The simple straight lines representing the transition metal and composite divides in Figure 9–3 are recognized to be oversimplifications in particular contrast to the ionic divide.

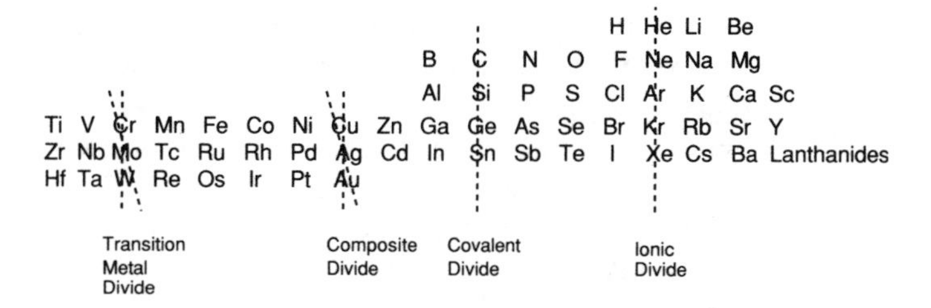

FIGURE 9–3. A major segment of the periodic table showing the four "divides" of H. E. M. Stone.

The "*s*-*d* shear" hypothesis[11] broadens the demarcation lines of these divides as indicated in Figure 9–3, so that these divides appear more to the right of the periodic table with the heavier elements. An experimental consequence of the *s*-*d* shear hypothesis occurs in the heats of atomization of the transition metals, which reflect the stability of the metal lattice and thus the strength of the overall metal-metal bonding in the bulk metal structure. In the 5*d* transition series, the heats of atomization have a maximum at the group 6 metal tungsten, but in the 3*d* and 4*d* series they have maxima at the group 5 metals vanadium and niobium rather than the corresponding group 6 metals chromium and molybdenum, respectively.

This model for the chemical bonding in metals and alloys may be considered to introduce topology-derived ideas to the interstitial electron model developed by Johnson in the early 1970's.[12,13,14,15,16,17] The Johnson model originates from the so-called Hellmann-Feynmann theorem and leads to the localization of itinerant electron density in octahedral or tetrahedral interstices of close-packed metal ion cores in metals and alloys. Johnson determines such electron occupancy by the requirements of minimum electron-electron repulsion and of opposite spins for electrons in adjacent interstices. Johnson has used this method to interpret numerous physical properties of metals, including the heat capacity, electrical conductivity, and magnetic susceptibility.

---

[11]H. E. N. Stone, *J. Mater. Sci.*, **12**, 201, 1977.

[12]O. Johnson, *Bull. Chem. Soc. Japan*, **45**, 1599, 1972).

[13]O. Johnson, *Bull. Chem. Soc. Japan*, **45**, 1607, 1972.

[14]O. Johnson, *Bull. Chem. Soc. Japan*, **46**, 1919, 1972.

[15]O. Johnson, *Bull. Chem. Soc. Japan*, **46**, 1923, 1972.

[16]O. Johnson, *Bull. Chem. Soc. Japan*, **46**, 1929, 1972.

[17]O. Johnson, *Bull. Chem. Soc. Japan*, **46**, 1935, 1972.

9.2 SUPERCONDUCTORS HAVING DIRECT METAL-METAL BONDING

Superconductors[18] are materials whose electrical resistance vanishes below a certain temperature, commonly called the *critical temperature* and designated as T_c. The first known superconductor was elemental mercury, whose superconducting properties were first discovered by Kammerlingh Onnes in 1911,[19] as indicated by the observation that the resistance of a sample of mercury dropped from 0.08 Ω at above 4 K to less than3 × 10^{-6} Ω at about 3 K and that this drop occurred over a temperature interval of only 0.01 K. The very low value of T_c, not only for elemental mercury but for essentially all other superconductors, has limited the practical applications of superconductors; a continuing objective of solid state physics research has been to find superconductors with higher T_c values so as to facilitate their practical application. Only in the case of the relatively recently discovered copper oxide superconductors[20] have materials been found which can be used above liquid nitrogen temperature, and the highest known T_c even in these copper oxide systems is still only ~120 K. Other physical characteristics of superconductors that influence their applications include the destruction of superconductivity even below T_c by application of a magnetic field larger than the so-called critical magnetic field H_c, the quenching of superconductivity by passing a *critical current* J_c through the superconductor, and the vanishing of the magnetic induction B inside the superconductor (the so-called *Meissner effect*). Measurements of the Meissner effect (i.e., the magnetic induction as the temperature is lowered) is often a more sensitive method of detecting superconductivity than measurement of the electrical resistance as the temperature is lowered, particularly in samples of superconductors contaminated with other non-superconducting materials.

The interest in developing superconductors with ever increasing T_c's for practical reasons has stimulated theoretical work as to the origin of superconductivity. The climax of this work was the theory of Bardeen, Cooper, and Schrieffer[21] in 1957. This theory modeled superconductivity by a gas of electrons interacting with each other through a two-particle interaction, which was shown to be the lattice vibrations ("phonons") on the basis of still earlier (1950) work by Fröhlich.[22] The pairs of *mobile* electrons responsible for superconductivity in a metallic structure are commonly called *Cooper pairs*. Note that the pairing of conduction electrons to form Cooper pairs leading to superconductivity is totally different than the

---

[18]R. D. Parks, Ed. *Superconductivity*, Dekker, New York, 1969.

[19]H. K. Onnes, *Leiden Comm.*, 1206, 1226, 1911, as quoted in D. Shoenberg, *Superconductivity*, Cambridge, New York, 1952, 1.

[20]R. M. Metzger, Ed., *High Temperature Superconductivity: The First Two Years*, Gordon and Breach, New York, 1988.

[21]J. Bardeen, L. N. Cooper, and J. R. Schrieffer, *Phys. Rev.*, **108**, 1175, 1957.

[22]H. Fröhlich, *Phys. Rev.*, **79**, 845, 1950.

pairing of valence electrons so familiar to chemists, and these two types of electron pairing should not be confused. The physical theories of superconductivity are rather complicated and certainly beyond the scope of this book, particularly since they provide relatively little insight as to what types of chemical structures make good superconductors. However, some of the graph-theory derived methods for studying chemical structure and bonding can be applied to superconductors based on their infinite solid state structures. This section treats such structures containing direct metal-metal bonds. The next chapter will treat analogously metal oxide superconductors, particularly the high T_c copper oxide superconductors, but only after discussing some general aspects of metal oxide systems.

The requirement of Cooper electron pairs for superconductivity can be related to the graph-theory derived electronic structure of bulk metals discussed in the previous section with regard to the effect of the average number of metal valence electrons. This theory indicates a special role for the group 6 metals chromium, molybdenum, and tungsten in forming the so-called "transition metal divide", since their six valence electrons are exactly enough for one multicenter bond in each octahedral cavity and in each of the two tetrahedral cavities associated with each octahedral cavity. This bonding topology corresponds to complete occupancy of these delocalized bonding orbitals with electron pairs (i.e., the analogue of a "closed shell" electronic configuration) and thus is very unfavorable for the *mobile* Cooper electron pairs required for superconductivity. This is consistent with the very low T_c's (< 0.1 K) observed for chromium, molybdenum, and tungsten.[23,24]

Now consider the group 5 metals vanadium, niobium, and tantalum, which have much higher T_c's than the group 6 metals considered above. The above bonding model for the group 5 metals leaves only a single electron rather than an electron pair for one of the three multicenter bonds associated with a given octahedral cavity (including the two tetrahedral cavities associated with each octahedral cavity). These single electrons can interact to form the mobile Cooper pairs required for superconductivity, accounting for the much higher T_c's for group 5 metals relative to the group 6 metals and the local maximum in the T_c versus Z_{av} curve at $Z_{av} = 4.8$ for transition metal alloys.[23,24] Similarly, for the group 7 metals technetium and rhenium, the above bonding model leaves an extra electron after providing electron pairs for each of the three multicenter bonds associated with a given octahedral cavity. Pairing of these extra electrons can lead to the mobile Cooper pairs required for superconductivity, thereby accounting for the much higher T_c's of group 7 metals relative to the group 6 metals and the local maximum in the T_c versus Z_{av} curve at $Z_{av} = 7$ for transition metal alloys. The smaller local maximum in the T_c versus Z_{av} curve at $Z_{av} = 3.3$ may have a similar origin, based on pairing of the single electron remaining from each metal atom after an

[23]B. T. Matthias, *Phys. Rev.*, **97**, 74, 1955.

[24]B. T. Matthias in *Progress in Low Temperature Physics II*, C. J. Gorter, Ed., North Holland, Amsterdam, 1975, 138.

electron pair is provided for each six-center core bond in the centers of the octahedral cavities with no multicenter bonds in the tetrahedral cavities.

A similar graph-theory derived approach can be used to study the chemical bonding topology of ternary superconductors, i.e., superconductors containing three different elements including at least one *d*-block transition metal. Such superconductors contain a *conducting skeleton* consisting of the transition metal subskeleton with extensive direct metal-metal bonding within this subskeleton. Of particular interest are the ternary molybdenum chalcogenides, often called Chevrel phases.[25,26] These phases were the first superconducting ternary structures found to have relatively high T_c's reaching 15 K for $PbMo_6S_8$,[27] in addition to having relatively high critical fields. From the structural point of view, the Chevrel phases are constructed from Mo_6 octahedra, which, depending upon the system, can be discrete (i.e., joined only at vertices) and/or fused together. The discrete Mo_6 octahedra in Chevrel phases may be considered to have a bonding topology analogous to that in the molybdenum(II) halides $Mo_6X_8L_6^{4+}$ (L = two-electron donor ligand) discussed in Section 7.1. Fusion of molybdenum octahedra in the Chevrel phases involves sharing of opposite triangular faces, similar to some rhodium carbonyl anions (Section 6.6 and Figure 6–10) but differing from the sharing of opposite edges found in the infinite chain lanthanide halide clusters such as Gd_2Cl_3 (Section 9.1 and Figure 9–1).

First, consider the Chevrel phases of the general formula $M_nMo_6S_8$ and $M_nMo_6Se_8$ (M = Ba, Sn, Pb, Ag, lanthanides, Fe, Co, Ni, etc.).[25] The basic building blocks of their structures are Mo_6S_8 (or Mo_6Se_8) units containing a bonded Mo_6 octahedron (Mo–Mo distances in the range 2.67 to 2.78 Å) with a sulfur atom capping each face, leading to an Mo_6 octahedron within an S_8 cube, the same as the Mo_6X_8 structural unit for the $Mo_6X_8L_6^{4+}$ halides depicted in Figure 7–1a. Each (neutral) sulfur atom of the S_8 cube functions as a donor of four skeletal electrons to the Mo_6 octahedron within that S_8 cube, leaving a sulfur electron pair to function as a ligand to a molybdenum atom in an adjacent Mo_6 octahedron. Maximizing this sulfur electron pair donation to the appropriate Mo_6 octahedron results in a tilting of the Mo_6 octahedron by about 25° within the cubic array of the other metal atoms M.[28] These other metal atoms M furnish electrons to these Mo_6S_8 units, allowing them to approach but not attain the $Mo_6S_8^{4-}$ closed-shell electronic configuration. This corresponds to the partially filled valence band of a conductor. Electronic bridges between individual Mo_6 octahedra are provided by interoctahedral metal-metal interactions (nearest *inter*octahedral Mo–Mo distances in the range 3.08 to 3.49 Å for Mo_6S_8 and Mo_6Se_8 derivatives.

---

[25]Ø. Fischer, *Appl. Phys.*, **16**, 1, 1978.

[26]R. Chevrel, P. Gougeon, M. Potel, and M. Sergent, *J. Solid State Chem.*, **57** 25, 1985.

[27]B. T. Matthias, M. Marezio, E. Corenzwit, A. S. Cooper, and H. E. Barz, *Science*, **175**, 1465 (1972).

[28]J. K. Burdett and J.-H. Lin, *Inorg. Chem.*, **21**, 5, 1982.

The $Mo_6S_8^{4-}$ closed-shell electronic configuration for the fundamental Chevrel phase building block is isoelectronic with that of the $Mo_6X_8L_6^{4+}$ halides discussed in Section 7.1, remembering that each molybdenum vertex receives an electron pair from a sulfur atom of an adjacent Mo_6S_8 unit and thus may be treated as an LMo vertex. This leads to the following electron counting scheme for the closed-shell $Mo_6S_8^{4-}$ unit:

6 LMo vertices: (6)(–2) =	–12 electrons
8 μ_3-S bridges: (8)(4) =	32 electrons
–4 charge:	4 electrons
Total skeletal electrons:	*24 electrons*

These 24 skeletal electrons are the exact number required for an edge-localized octahedron having two-electron two-center bonds along each of its 12 edges.

This chemical bonding topology of the Chevrel phases MMo_6S_8, consisting of edge-localized discrete Mo_6 octahedra linked through sulfur atoms as well as interoctahedral metal-metal interactions, leads naturally to the concept which has been called[29] *porous delocalization*. Thus, the bonding in a network of polyhedra with edge-localized bonding is porous in contrast to the dense bonding in a network of polyhedra with globally delocalized bonding. In other words, a porous chemical bonding topology uses only the 1-skeleton[30] of the polyhedron, in contrast to the dense bonding in a polyhedron which uses the whole volume of the polyhedron. It has been conjectured[29] that a porously delocalized three-dimensional network consisting of electronically linked polyhedral metal clusters having *edge-localized* chemical bonding leads to superconductors having relatively high critical fields and temperatures. According to this conjecture, the porosity of the chemical bonding in the MMo_6S_8 Chevrel phases makes their superconductivity less susceptible to magnetic fields and temperature than that of densely delocalized structures such as pure metals. This idea also has been related to the suggestion[31] that the high critical fields of the Chevrel phases MMo_6S_8 arise from a certain localization of the conduction-electron wavefunction of the Mo_6 clusters, leading to an extremely short mean free path and/or a low Fermi velocity corresponding to a small coherence length in the Bardeen/Cooper/Schrieffer theory mentioned above. Further details of these ideas are beyond the scope of this book.

The Chevrel phases include not only the species constructed from discrete Mo_6S_8 (or Mo_6Se_8) octahedra but also species constructed from Mo_9S_{11}, $Mo_{12}S_{14}$, and $[Mo_6S_6]_\infty$ units, formed by the fusion of octahedra by sharing triangular faces (Figure 9–4). This fusion process may be regarded as analogous to the formation of polycyclic aromatic hydrocarbons by the fusion

---

[29]R. B. King, *J. Solid State Chem.*, **71**, 224, 1987.

[30]B. Grünbaum, *Convex Polytopes*, Interscience, New York, 1967.

[31]Ø. Fischer, M. Decroux, R. Chevrel, and M. Sergent in *Superconductivity in* d- *and* f-*Band Metals*, D. H. Douglas, Ed., Plenum Press, New York, 1976, 176.

of hexagons by sharing edges. Thus, fused molybdenum octahedra can be classified by the trivial name of the polycyclic benzenoid hydrocarbon having an analogous configuration of its planar hexagon building blocks (Figure 9–4), similar to the classification of rhodium carbonyl anion clusters having related structures discussed in Section 6.6.

The molybdenum atoms in the fused octahedra of Figure 9–4 are of two types, inner and outer. The outer molybdenum atoms (unstarred in Figure 9–4) are similar to those in the discrete octahedral Mo_6S_8 building blocks discussed above. They thus use four internal orbitals and receive an electron pair from a sulfur atom of an adjacent metal cluster unit (indicated by arrows in Figure 9–4). The inner molybdenum atoms (unstarred in Figure 9–4) use six internal orbitals and do *not* receive an electron pair from a sulfur atom of an adjacent metal cluster. They are thus zero electron donors $[(3)(2) - 6 = 0]$. Edges connecting pairs of inner molybdenum atoms are bridged by sulfur atoms, but these sulfur atoms also bond to one molybdenum atom in each adjacent Mo_3 triangle, so that they function as pseudo five-coordinate μ_4 sulfur atoms and donors of four skeletal electrons to their own cluster units. Thus, all sulfur atoms in the species depicted in Figure 9–4 function as four-electron donors when considered as neutral ligands to a single aggregate of face-fused Mo_6 octahedra. The electron and orbital counting of these structures can then proceed as follows, considering only orbitals involved in the metal-metal bonding:

1. Naphthalene analogue, $Mo_9S_{11}^{4-}$:

Source of skeletal orbitals and electrons:

6 outer LMo vertices	–12 electrons	24 orbitals
3 inner Mo vertices	0 electrons	18 orbitals
11 S atoms: (11)(4) =	44 electrons	
–4 charge:	4 electrons	
Total available skeletal electrons and orbitals:	*36 electrons*	*42 orbitals*

Use of skeletal electrons and orbitals for metal-metal bonding of various types:

6 edge bonds on outer triangles	12 electrons	12 orbitals
6 face bonds connecting inner and outer triangles	12 electrons	18 orbitals
6 edge bonds connecting inner and outer triangles	12 electrons	12 orbitals
Total skeletal electrons and orbitals used:	*36 electrons*	*42 orbitals*

Note that the chemical bonding topology of $Mo_9S_{11}^{4-}$ contains three three-center Mo–Mo–Mo face bonds in each of the two octahedra between an outer and an inner Mo_3 triangle, similar to the three-center Nb–Nb–Nb face bonds in the $Nb_6X_{12}L_6^{2+}$ discussed in Section 7.1.

2. Anthracene analogue, $Mo_{12}S_{14}^{6-}$:

Source of skeletal orbitals and electrons:

6 outer LMo vertices	–12 electrons	24 orbitals
6 inner Mo vertices	0 electrons	36 orbitals
14 S atoms: (14)(4) =	56 electrons	
–6 charge	6 electrons	
Total available skeletal electrons and orbitals:	*50 electrons*	*60 orbitals*

Use of skeletal orbitals and electrons for metal-metal bonding of various types:

Globally delocalized central Mo_6 octahedron:		
Surface bonding:	12 electrons	12 orbitals
Core bonding:	2 electrons	6 orbitals
Outer Mo_3 triangles:		
6 edge bonds:	12 electrons	12 orbitals
Octahedra formed by an outer and an inner Mo_3 triangle:		
6 face bonds:	12 electrons	18 orbitals
6 edge bonds:	12 electrons	12 orbitals
Total skeletal electrons and orbitals used:	*50 electrons*	*60 orbitals*

This bonding model suggests that, of the three octahedral cavities in the $Mo_{12}S_{14}^{6-}$ structure, the octahedral cavity formed by the two inner Mo_3 triangles has globally delocalized bonding, whereas the two equivalent octahedral cavities formed by one outer and one inner Mo_3 triangle have the same combination of two-center (edge) and three-center (face) bonds as the two (equivalent) octahedral cavities in the $Mo_9S_{11}^{4-}$ naphthalene analogue cluster discussed above.

Face-sharing fusion of molybdenum octahedra can be continued to give in the infinite limit the linear polyacene analogues $[Mo_6S_6^{2-}]_\infty$ (Figure 9–4) known in a number of derivatives $[M_2Mo_6S_6]_\infty$ (M = K, Rb, Cs), as well as their selenium analogues $[M_2Mo_6Se_6]_\infty$ (M = Na, K, Rb, Cs, Tl, Ag) and the tellurium analogues $[M_2Mo_6Te_6]_\infty$ (M = Rb, Cs, In, Tl).[32,33] The

[32]M. Potel, R. Chevrel, M. Sergent, J. C. Armici, M. Decroux, and Ø. Fischer, *J. Solid State Chem.*, **35**, 286, 1980.

[33]P. H. Hor, W. C. Fan, L. S. Chou, R. L. Meng, C. W. Chu, J. M. Tarascon, and M. K. Wu, *Solid State Comm.*, **55**, 231, 1985.

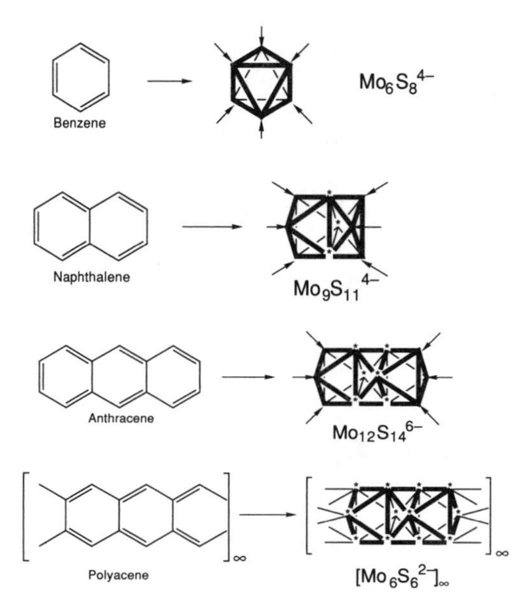

FIGURE 9–4. Analogy between the fusion of molybdenum octahedra in ternary molybdenum sulfide structures and the fusion of benzene rings in planar polycyclic hydrocarbons. Unstarred vertices are outer molybdenum atoms, and starred vertices are inner molybdenum atoms. Arrows indicate sites of coordination with sulfur atoms of adjacent metal cluster units. Sulfur atoms are omitted for clarity. Similar structural units are present in analogous molybdenum selenides and tellurides.

structures of these systems consist of infinite chains of face-fused metal octahedra as depicted in Figure 9–4. All molybdenum atoms are inner molybdenum atoms (starred in Figure 9–4), and none of the chalcogens bridge to other chains so that there are no close contacts between the different chains. In accord with this structure these systems function as pseudo-one-dimensional metals having strongly anisotropic conductivities several hundred times larger parallel to the chains of octahedra relative to the perpendicular directions.[34] The $Mo_{6/2}S_{6/2}^-$ octahedra serving as building blocks for these $[M_2Mo_6S_6]_\infty$ derivatives, and their selenium and tellurium analogues have 13 skeletal electrons, i.e., none from the (inner) molybdenum vertices, 12 (= (3)(4)) from the three (= 6/2) sulfur atoms, and 1 from the –1 charge. These 13 skeletal electrons per $Mo_{6/2}S_{6/2}^-$ unit are one less than the 14 skeletal electrons required for the octahedral cavity to be globally delocalized ($2n + 2 = 14$ for $n = 6$). These holes in the closed-shell electronic configurations for globally delocalized $[Mo_6S_6^{2-}]_\infty$ correspond to holes in their valence bands and provide a mechanism for electronic conduction along the chains of face-linked octahedra.

---

[34] J. C. Armici, M. Decroux, Ø. Fischer, M. Potel, R. Chevrel, and M. Sergent, *Solid State Comm.*, **33**, 607, 1980.

The conjecture, suggested by the analysis outlined above of the chemical bonding topology of the Chevrel phases MMo_6E_8 (E = S, Se, Te), that porous infinite delocalization is a feature of the chemical bonding topology of superconductors exhibiting relatively high critical temperatures and critical magnetic fields is supported by an analysis of the chemical bonding topologies of other ternary superconductors. Particularly clear cases are the ternary lanthanide rhodium borides, $LnRh_4B_4$ (Ln = certain lanthanides such as Nd, Sm, Er, Tm, Lu), another class of relatively high-temperature non-oxide superconductors[35,36] exhibiting significantly higher T_c's than other types of metal borides. These rhodium borides have structures consisting of electronically linked discrete Rh_4 tetrahedra. The topology of an individual Rh_4B_4 unit in these ternary borides is that of a tetracapped tetrahedron of T_d symmetry, in which the four degree 6 vertices correspond to rhodium atoms and the four degree 3 vertices correspond to boron atoms. Such a tetracapped tetrahedron is topologically equivalent to a cube with six diagonals drawn to preserve T_d overall symmetry (Figure 9–5). The face diagonals of the Rh_4B_4 cube in the $LnRh_4B_4$ borides correspond to six Rh–Rh bonds (average length 2.71 Å in YRh_4B_4), and the edges of the cube correspond to 12 Rh–B bonds (average length 2.17 Å in YRh_4B_4).[37] The ratio between these two lengths, namely 2.71/2.17 = 1.25, is only about 13% less than the $\sqrt{2}$ = 1.414 ratio of these lengths in an ideal cube, suggesting that the Rh_4B_4 building blocks can be approximated by a cube in the three-dimensional lattice. The Rh–Rh distances of 2.71 Å in these Rh_4B_4 units are essentially identical to the mean Rh–Rh distance in the discrete molecular tetrahedral rhodium cluster[38] $Rh_4(CO)_{12}$, regarded as a prototypical example of an edge-localized tetrahedron.

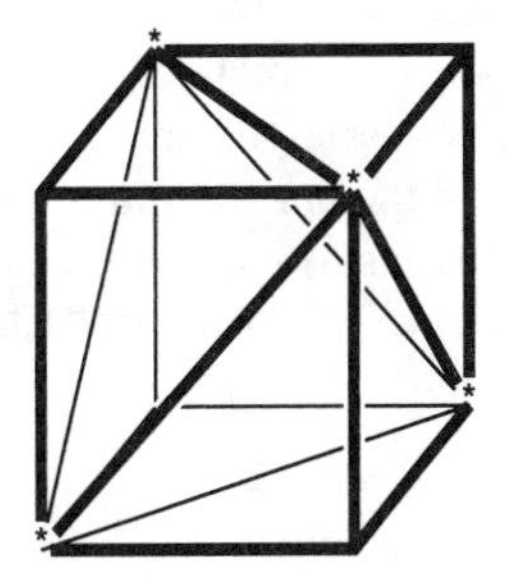

FIGURE 9–5. A cube with six diagonals which is topologically equivalent to a tetracapped tetrahedron. The degree 6 vertices are starred for clarity; these vertices correspond to the vertices of the tetrahedron.

---

[35] B. T. Matthias, E. Corenzwit, J. M. Vandenberg, and H. E. Barz, *Proc. Natl. Acad. Sci. U.S.A.*, **74**, 1334, 1977.

[36] L. D. Woolf, D. C. Johnston, H. B. MacKay, R. W. McCallum, and M. B. Maple, *J. Low Temp. Phys.*, **35**, 615, 1979.

[37] J. M. Vandenberg and B. T. Matthias, *Proc. Natl. Acad. Sci. U.S.A.*, **74**, 1336, 1977.

[38] F. H. Carré, F. A. Cotton, and B. A. Frenz, *Inorg. Chem..*, **15**, 380, 1976.

The arrangement of the Rh_4B_4 units as sheets in the primitive tetragonal lattice of $LnRh_4B_4$ is shown in Figure 9–6. All of the valence orbitals of both the boron and rhodium atoms are involved in the formation of four and nine, respectively, two-center bonds. The four bonds formed by a boron atom using its sp^3 bonding orbital manifold are as follows:

1. Three bonds to rhodium atoms in the same Rh_4B_4 cube (average Rh–B distance 2.17 Å in YRh_4B_4).
2. One bond to the nearest boron atom in an adjacent Rh_4B_4 cube (B–B distance 1.86 Å in YRh_4B_4), thereby leading to discrete B_2 units in the structure.

The nine bonds formed by a rhodium atom using its sp^3d^5 bonding orbital manifold are as follows:

1. Three bonds to rhodium atoms in the same Rh_4 tetrahedron (average Rh–Rh distance 2.71 Å in YRh_4B_4).
2. Three bonds to boron atoms in the same Rh_4B_4 cube (average Rh–B distance 2.17 Å in YRh_4B_4).
3. One bond to the nearest rhodium atom in another Rh_4B_4 cube in the same sheet (Rh–Rh distance 2.68 Å in YRh_4B_4).
4. Two bonds to the next-nearest rhodium atoms in adjacent Rh_4B_4 cubes (Rh–Rh distances 3.14 Å in YRh_4B_4).

In deriving the chemical bonding topology, each boron atom is considered to have three internal orbitals and one external orbital and is therefore a donor of two skeletal electrons, since one of the three boron valence electrons is needed for the B–B bond using its external orbital. Similarly, each rhodium atom has six internal orbitals and three external orbitals and is therefore a donor of six skeletal electrons, since three of the nine rhodium valence electrons are needed for the three external Rh–Rh bonds formed by given rhodium atoms.

These considerations indicate that a neutral Rh_4B_4 unit in the $LnRh_4B_4$ borides has 32 skeletal electrons as follows:

4 Rh vertices: (4)(6) =	24 electrons
4 B vertices: (4)(2) =	8 electrons
Total skeletal electrons for each Rh_4B_4 unit:	*32 electrons*

Since a tetracapped tetrahedron or the topologically equivalent cube with six diagonals (Figure 9–5) has 18 edges, corresponding to six Rh–Rh bonds and 12 Rh–B bonds as outlined above, a closed-shell edge-localized Rh_4B_4 unit requires (2)(18) = 36 skeletal electrons, corresponding to the tetraanion $Rh_4B_4^{4-}$. Since the lanthanides also present in the lattice form tripositive rather than tetrapositive ions, the $LnRh_4B_4$ borides must be $Ln^{3+}Rh_4B_4$

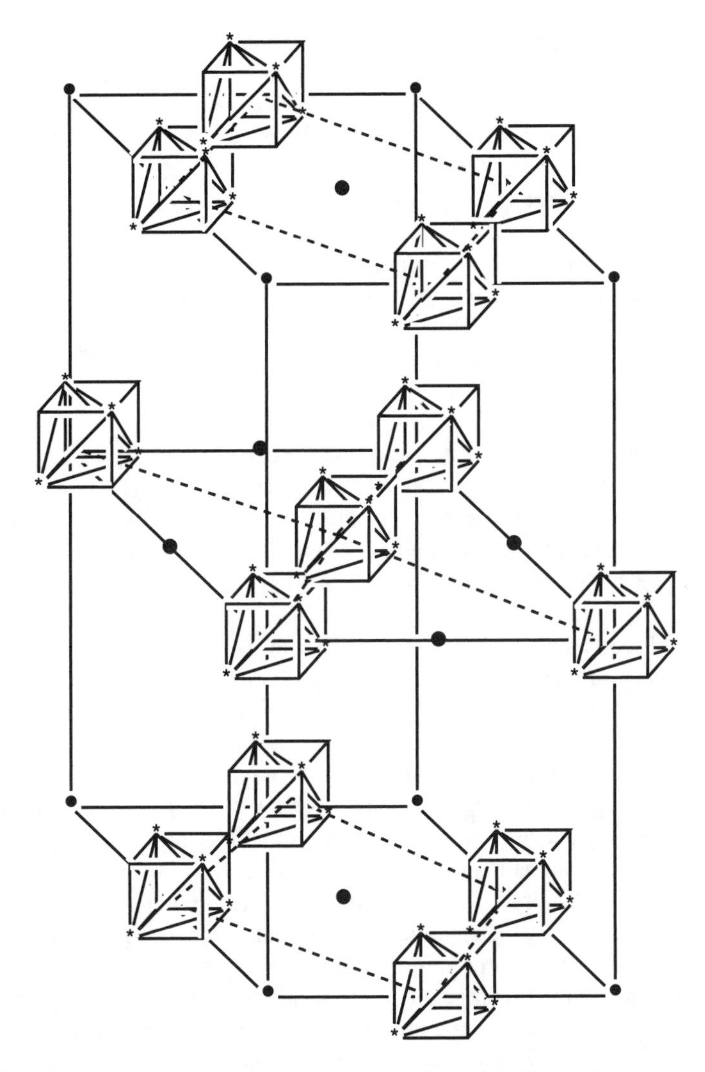

FIGURE 9–6. The primitive tetragonal structure of $LnRh_4B_4$. The solid circles represent the lanthanide atoms. The cubes, not drawn to scale, represent the Rh_4B_4 units; the rhodium vertices are starred and individual Rh–Rh and B–B bonds are omitted for clarity. The dotted lines connecting the Rh_4B_4 cubes correspond to one B–B bond (~1.86 Å), one Rh–Rh bond (~2.68 Å), and two Rh–Rh bonds (~3.14 Å).

with the $Rh_4B_4^{3-}$ anion having one electron less than the closed shellelectronic configuration $Rh_4B_4^{4-}$. This situation is similar to that discussed above for the Chevrel phases, in which, for example, $PbMo_6S_8$ has $Mo_6S_8^{2-}$ units with two electrons less than the closed-shell electronic configuration $Mo_6S_8^{4-}$, and the infinite chain conductors $M^I_2Mo_6S_6$ have octahedral $Mo_{6/2}S_{6/2}^-$ units with one electron less than the closed-shell electronic configuration $Mo_{6/2}S_{6/2}^{2-}$. These electron deficiencies appear to be an important feature of these highly conducting systems, since they provide holes in the valence bands for the electron mobility required for conductivity.

The chemical bonding topologies of other ternary metal boride superconductors, such as $M^{II}_{0.67}Pt_3B_2$, $LnRuB_2$, and $LnOs_3B_2$, can be treated by analogous methods.[39] In all cases, the conducting skeleton consists of an infinite metal network constructed largely from two-electron two-center edge-localized bonds.

Another important class of superconducting materials are the so-called A-15 alloys. Before the discovery of the high T_c copper oxide superconductors, the highest T_c's were found in alloys having the A-15 structure and the stoichiometry M_3E, in which M is Nb or V and E is Ga, Si, Ge, or Sn.[40,41] Approaches analogous to those described above can be used to treat the skeletal bonding topology of the A-15 alloys.

Consider V_3Si as a prototypical A-15 alloy, which may be regarded as isoelectronic with the important A-15 superconductors Nb_3Ge and Nb_3Sn. Figure 9–7 depicts a cubic building block $V_{12/2}SiSi_{8/8} = V_6Si_2$ of the A-15 structure, in which the silicon and vanadium atoms are represented by shaded and open circles, respectively. The structure has the following features:

1. A silicon atom is located in the center of the cube (= Si).
2. Silicon atoms are located at the eight vertices of the cube, which are each shared with seven other cubes (= $Si_{8/8}$);
3. Pairs of vanadium atoms are located in each of the six faces of the cube. Each vanadium pair is shared with the adjacent cube sharing the same face (= $V_{12/2}$). The open circles indicating the pairs of vanadium atoms in a given face are connected by lines for clarity in Figure 9–7;
4. The vanadium pairs in the six faces are oriented to form infinite straight vanadium chains in each of the three orthogonal directions. These chains comprise the essential portion of the conducting skeleton;
5. The polyhedron formed by the 12 vanadium atoms around the central silicon atom is topologically an icosahedron having 20 triangular faces, eight of which are capped by the vertex silicon atoms, thereby forming tetrahedral V_3Si cavities. The 20 – 8 = 12 uncapped faces of the V_{12} icosahedron occur in six "diamond pairs" of the following type:

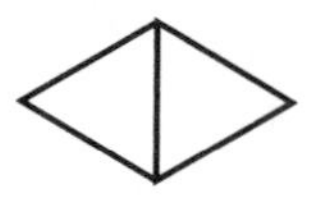

Consider a chemical bonding topology for V_3Si in which there are two-electron two-center V–V bonds in the infinite straight vanadium chains in each of the three directions. Six such vanadium chains touch a given V_6Si_2

[39] R. B. King, *Inorg. Chem.*, **29**, 2164, 1990.
[40] J. M. Leger, *J. Low Temp. Phys.*, **14**, 297, 1974.
[41] G. R. Johnson and D. H. Douglass, *J. Low Temp. Phys.*, **14**, 575, 1974.

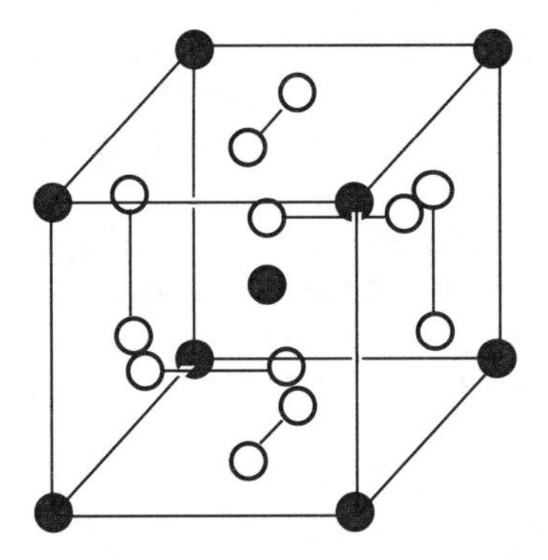

FIGURE 9–7. The V_6Si_2 cubic building block for V_3Si and related A-15 superconductors. Shaded circles in the center and at each of the eight vertices correspond to silicon atoms, whereas the pairs of open circles (each linked by a straight line) in each of the six faces correspond to vanadium atoms.

reference cube (Figure 9–7), namely one in each face. The V–V bonds of the vanadium chains, although equivalent in the overall structure, are of the following two types relative to a V_6Si_2 reference cube with which they are associated:

1. The first type (A) consists of bonds between two vanadium atoms in a given face of the V_6Si_2 reference cube. Such V–V bonds are shared only between the two cubes sharing the face containing the vanadium atoms.
2. The second type (B) consists of bonds between a vanadium atom and the adjacent vanadium atom in the chain *not* associated with the V_6Si_2 reference cube. Such V–V bonds are shared between four adjacent V_6Si_2 cubes.

A given V_6Si_2 cube has six V–V bonds of type A and 12 V–V bonds of type B. Thus the V–V bonds associated with a given V_6Si_2 reference cube require $(2)(6/2 + 12/4) = 12$ skeletal electrons and use an equivalent number of vanadium valence orbitals, since two-electron two-center bonds are being considered.

The chemical bonding topology for V_3Si summarized in Table 9–1 can now be defined relative to a V_6Si_2 reference cube. Since each vanadium atom uses nine valence orbitals (i.e., an sp^3d^5 valence orbital manifold) and each silicon atom uses four valence orbitals (i.e., an sp^3 valence orbital manifold), the 54 ($= 9 \times 6$) vanadium valence orbitals and eight ($= 2 \times 4$) silicon valence orbitals required for the bonding topology described in Table 9–1 correspond exactly to what is available in a V_6Si_2 cube. A neutral V_6Si_2 cube has $(6)(5) + (2)(4) = 38$ valence electrons, which is two less than the 40 valence electrons required for the chemical bonding topology outlined in Table 9–1, implying that $V_6Si_2^{2-}$ is the closed-shell electronic configuration for this chemical bonding topology. This model for the chemical bonding in the A-15 superconductors V_3Si and its analogues Nb_3Ge and Nb_3Sn thus has the following features associated with superconductivity as discussed above:

1. A porous conducting skeleton in which straight chains of V–V edge-localized bonds are an important component;
2. Holes in the valence band, since $V_6Si_2^{2-}$ rather than V_6Si_2 is the closed-shell electronic configuration.

TABLE 9–1
Chemical Bonding Topology of V_3Si and Related A-15 Superconductors Such as Nb_3Ge and Nb_3Sn

Bond type	Number of such bonds	Number of electrons required	Number of V orbitals required	Number of Si orbitals required
2-center V–V in chains	6	12	12	0
4-center V_3Si in V_3Si tetrahedral cavities	8	16	24	8
3-center V_3 in one V_3 face of each diamond pair	6	12	18	0
Total electrons and orbitals required for one V_6Si_2 cube:		*40*	*54*	*8*

Chapter 10

METAL OXIDES WITH METAL-METAL INTERACTIONS

10.1 METAL-METAL INTERACTIONS THROUGH OXYGEN ATOMS

The metal clusters discussed in Chapters 6 through 9 have direct chemical bonds between metal atoms. This chapter discusses metal oxides in which there are indirect metal-metal interactions through oxygen atoms. Reduced polyoxometalates[1] of metals such as molybdenum and tungsten form an important class of polyoxometalates in which there are metal-metal interactions through oxygen atoms. In addition, copper-copper interactions through oxygen atoms are found in the high critical temperature copper oxide superconductors.

Polynuclear transition metal compounds in which there are no direct metal-metal bonds are conveniently dissected into a collection of mononuclear components linked by bridging groups. In the species of interest in this chapter, the bridging groups are oxygen atoms. If some of the mononuclear components have unpaired electrons, then there are possibilities for indirect interactions in which the unpaired electron spins may become paired (antiferromagnetism) or may become parallel (ferromagnetism). Transition metal derivatives with the potential for such indirect metal-metal interactions between paramagnetic transition metal atoms have been categorized according to their magnetic behavior into three main groups depending on the strength of the metal-metal interaction.[2] In the *noninteracting* type, the metal-metal interaction is negligible, so that the magnetic properties of the polynuclear derivative are essentially unchanged from those of the paramagnetic mononuclear derivative. In the *strongly interacting* type, relatively strong direct metal-metal bonds are formed so that the polynuclear derivative displays simple diamagnetic behavior. This chapter is concerned with the intermediate case, namely *weakly interacting* metal ions. The weak coupling between the unpaired electrons of the individual metal atoms in such polynuclear derivatives leads to low-lying excited states which can be populated at thermal energies (≤ 1000 cm^{-1}). Such interactions are sometimes called *superexchange* because of the relatively large distances (3 to 5 Å) between the metal ions. Superexchange is particularly prevalent in polynuclear derivatives constructed from mononuclear components in which the central transition metal has one electron either more or less than a closed shell electronic configuration such

---

[1]M. T. Pope, *Heteropoly and Isopoly Oxometalates*, Springer-Verlag, Berlin, 1983.

[2]P. J. Hay, J. C. Thibeault, and R. Hoffmann, *J. Am. Chem. Soc.*, **97**, 4884, 1975.

as d^1 Ti(III) and d^9 Cu(II) and in which the geometry of the bridging groups forces the interacting metal atoms to remain at a non-bonding distance.

The first theoretical work on the energetics of coupled metal pairs was published by Anderson,[3] who used a valence bond formalism. Schematically for a d^1–d^1 (or d^9–d^9) system, the unpaired electron on each metal center is assigned to an orbital, called a *magnetic orbital.* Let ϕ_A and ϕ_B be the wave functions of the magnetic orbitals of atoms A and B respectively. In the absence of any metal-metal interaction, the unpaired electron around any given metal atom is centered on the metal but partially delocalized towards any bridging and/or terminal ligands surrounding it. The interaction between the two metals leads to two molecular states: a spin singlet and triplet separated by energy J. If relatively few assumptions are made, this singlet-triplet gap can be expressed as a sum of two components: a negative antiferromagnetic term J_{AF} and a positive ferromagnetic contribution J_F. Thus

$$J = J_{AF} + J_F \tag{10–1}$$

For a non-symmetrical d^1–d^1 (or d^9–d^9) system

$$J_{AF} = -2S\sqrt{(\Delta^2 - \delta^2)} \tag{10–2a}$$

$$J_F = 2j \tag{10–2b}$$

where S is the overlap integral (see Section 4.1) between two magnetic orbitals, δ is the energy gap between the magnetic orbitals on atoms A and B, Δ is the energy gap between the two molecular orbitals constructed from the magnetic orbitals for the triplet state, and j is the two-electron exchange integral. If the overlap density $\rho(i)$ between these two magnetic orbitals is defined as $\rho(i) = \phi_A(i)\phi_B(i)$, then S and j are given by the equations:

$$S = \int_{\text{space}} \rho \, d\tau \tag{10–3a}$$

$$j = \int_{\text{space}} \frac{\rho(i)\rho(j)}{r_{ij}} \, d\tau_i d\tau_j \tag{10–3b}$$

If metals A and B are the same, then $\delta = 0$ and $J_{AF} = -2\Delta S$.

The simplest situation in the above case occurs when the relative orientation of the magnetic orbitals on metals A and B is unfavorable towards any interaction; i.e., ρ is negligible and J_{AF} and J_F are both zero. In this case,

---

[3]P. W. Anderson, *Phys. Rev.*, **115**, 2, 1959.

the magnetic properties of the dimer are simply those of the separate mononuclear fragments.

Another relatively simple case occurs when the interacting magnetic orbitals are orthogonal corresponding to a ferromagnetic system. In this case $S = 0$ although $\rho \neq 0$. From Equation 10–2a, $J_{AF} = 0$, and the experimentally observed J consists entirely of J_F, the ferromagnetic component. There are two types of orthogonality of magnetic orbitals, namely *strict* orthogonality imposed by the symmetry of the system and *accidental* orthogonality not directly related to symmetry.

The most common situation occurs when the J_{AF} component predominates, leading to antiferromagnetic behavior. Such behavior is particularly prevalent in the case of bridged copper(II) complexes,[4] which can serve as models for the high T_c copper oxides. The copper magnetic orbitals are the $d(x^2–y^2)$ orbitals, which have the unpaired electron in the d^9 electronic configuration. Interaction between the $d(x^2–y^2)$ orbitals of two copper atoms can occur through a p orbital on a bridging oxygen atom (Figure 10–1). Details of the orbital interactions in metal dimer complexes have been described by Hay, Thibeault, and Hoffmann.[2]

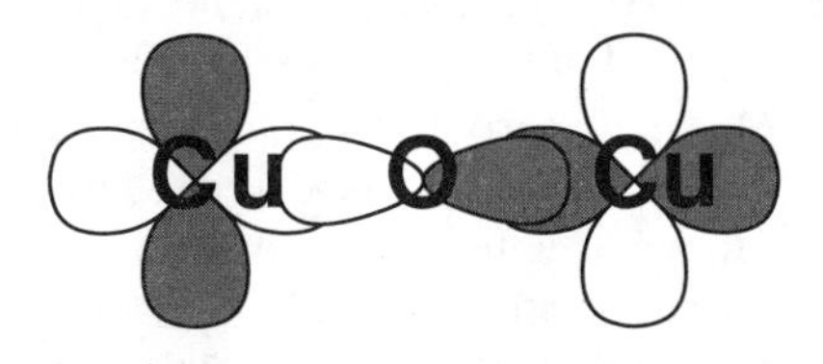

FIGURE 10–1. Interaction between two $d(x^2–y^2)$ copper orbitals through an oxygen p orbital.

10.2 DELOCALIZATION IN EARLY TRANSITION METAL POLYOXOMETALATES: BINODAL ORBITAL AROMATICITY

The heteropoly- and isopolyoxometalates of early transition metals[1,5] have been known for more than a century and have been studied extensively. Their structures are characterized by networks of MO_6 octahedra in which the early transition metals M (typically M = V, Nb, Mo, W) appear in their highest oxidation states in which they have a d^0 configuration. A characteristic of many, but not all, of such structures is their reducibility to highly colored mixed oxidation state derivatives, often given the trivial names of molybdenum or tungsten "blues." The reducibility of early transition metal

---

[4] R. J. Doedens, *Progr. Inorg. Chem.*, **21**, 209, 1976.

[5] V. W. Day and W. G. Klemperer, *Science*, **228**, 533, 1985.

polyoxometalates requires the presence of MO_6 octahedra in which only one of the six oxygen atoms is a terminal oxygen atom.[6] Such an MO_6 octahedron can be related to mononuclear L_5MO species[7] in which there is an essentially nonbonding metal *d* orbital to receive one or two electrons somewhat analogous to the copper $d(x^2-y^2)$ orbital depicted in Figure 10–1. The reducibility of the originally d^0 polyoxometalates can be related to the delocalization of the added electrons(s) in molecular orbitals formed by interaction of these nonbonding *d* orbitals on each of the metal atoms in the MO_6 octahedra forming the polyoxometalate structure. Such an approach was used by Nomiya and Miwa,[8] in the form of a structural stability index based on interpenetrating loops of the type –O–M–O–M–O– around the polyoxometalate cage, and suggested the analogy of closed loops of this type to macrocyclic π-bonding systems. The ready one-electron reducibility of a colorless to yellow polyoxomolybdate or polyoxotungstate to a highly colored mixed valence "blue" may be viewed as analogous to the one-electron reduction of benzenoid hydrocarbons such as naphthalene or anthracene to the highly colored corresponding radical anion. The graph-theory derived methods outlined in Chapter 4 for the study of delocalization in hydrocarbons and boranes can also be used to study delocalization in polyoxometalates.[9]

The polyoxometalates of interest consist of closed networks of MO_6 octahedra in which M is a d^0 early transition metal such as V(V), Nb(V), Mo(VI), or W(VI). These networks may be described by the large polyhedron or *macropolyhedron* formed by the metal atoms as vertices. In general, the edges of this macropolyhedron are M–O–M bridges, and with rare exceptions there is no direct metal-metal (M–M) bonding. The metal-metal interactions in the polyoxometalates are thus indirect metal-metal interactions through bridging oxygen atoms as discussed in Section 10.1 and similar to the indirect Cu–O–Cu interaction depicted in Figure 10–1.

The oxygen atoms in the polyoxometalates are of the following three types:

1. O^t: terminal or external oxygen atoms, which are multiply bonded to the metal (one σ and up to two orthogonal π bonds) and directed away from the macropolyhedral surface;
2. O^b: bridging or surface oxygen atoms, which form some or all of the macropolyhedral edges;
3. O^i: internal oxygen atoms, which are directed towards the center of the macropolyhedron.

---

[6]M. T. Pope, *Inorg. Chem.*, **11**, 1973 1972.

[7]C. J. Ballhausen and H. B. Gray, *Inorg. Chem.*, **1**, 111, 1962.

[8]K. Nomiya and M. Miwa, *Polyhedron*, **3**, 341, 1984.

[9]R. B. King, *Inorg. Chem.*, **30**, 4437, 1991.

The metal vertices of the macropolyhedron may be classified as $(\mu_n\text{-O})_5MO$ or *cis*-$(\mu_n\text{-O})_4MO_2$ vertices, depending upon the number and locations of the oxygen atoms. In the *cis*-$(\mu_n\text{-O})_4MO_2$ vertices, all nine orbitals of the sp^3d^5 manifold of M are used for the σ and π bonding to the two terminal oxygen atoms and σ bonding to the four bridging and internal oxygen atoms, leaving no orbitals for direct or indirect overlap with other metal vertices of the metal macropolyhedron, corresponding to a resonance hybrid depicted schematically in Figure 10–2a. The *cis*-$(\mu_n\text{-O})_4MO_2$ vertices in polyoxometalates correspond to the saturated CH_2 vertices in cyclohexane and other cycloalkanes. In the $(\mu_n\text{-O})_5MO$ vertices, only eight of the nine orbitals of the sp^3d^5 manifold of M can be used for σ and π bonding to the single oxygen atom and σ bonding to the five bridging and internal oxygen atoms leaving one nonbonding *d* orbital (d_{xy} if the M≡O(terminal) axis is the *z* axis) as depicted in Figure 10–2b. Thus, a $(\mu_n\text{-O})_5MO$ vertex with a nominally nonbonding d_{xy} orbital in a polyoxometalate is analogous to an unsaturated CH vertex with a nonbonding *p* orbital in a planar aromatic hydrocarbon such as benzene.

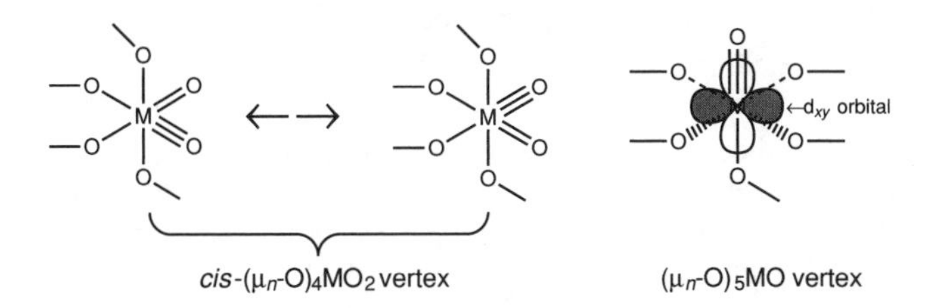

FIGURE 10–2. The metal-oxygen bonds and nonbonding atomic orbitals in *cis*-$(\mu_n\text{-O})_4MO_2$ and $(\mu_n\text{-O})_5MO$ vertices in polyoxometalates.

The nonbonding d_{xy} orbitals of the $(\mu_n\text{-O})_5MO$ vertices in the reducible early transition metal polyoxometalates have two orthogonal nodes (see Figure 10–2) and thus have improper four-fold symmetry (Chapter 2). Matching this four-fold orbital symmetry with the overall macropolyhedral symmetry requires macropolyhedra in which a C_4 axis passes through each vertex. A true three-dimensional polyhedron having C_4 axes passing through each vertex can have only O or O_h symmetry (the only point groups with multiple C_4 axes). The only two polyhedra having less than 15 vertices meeting these highly restrictive conditions are the regular octahedron and the cuboctahedron (Figure 10–3). It is therefore not surprising that these two polyhedra form the basis of the specific early transition-metal polyoxometalate structures containing only $(\mu_n\text{-O})_5MO$ vertices. Pope[1,6] calls such readily reducible structures containing only $(\mu_n\text{-O})_5MO$ vertices *type I structures*.

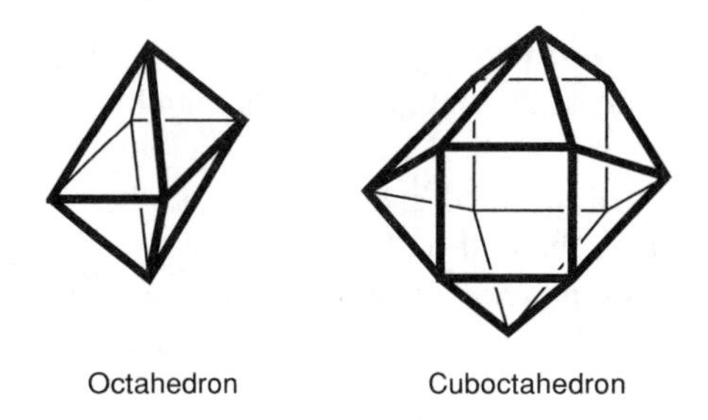

FIGURE 10–3. The octahedron and cuboctahedron found in polyoxometalate macropolyhedra.

The specific building blocks for type I polyoxometalate structures of interest are as follows:

1. **Octahedron:** $(MO^tO^b_{4/2}O^i_{1/6})_6 = M_6O_{19}^{n-}$ ($n = 8$, M = Nb, Ta; $n = 2$, M = Mo):
   O^t = one terminal oxygen atom per metal atom;
   $O^b_{4/2}$ = one bridging oxygen atom along each of the 12 edges of the octahedron (i.e., (4/2)(6) = 12);
   $O^i_{1/6}$ = a single μ_6 oxygen in the center of the M_6 macrooctahedron shared equally among all six metal vertices.
2. **Cuboctahedron (the so-called "Keggin Structure"):** $(MO^tO^b_{4/2}O^i_{1/3})_{12}X^{n-} = XM_{12}O_{40}^{n-}$ (n = 3 to 7; M = Mo, W; X = B, Si, Ge, P, Fe^{III}, Co^{II}, Cu^{II}, etc.)
   O^t = one terminal oxygen per metal atom;
   $O^b_{4/2}$ = one bridging oxygen along each of the 24 edges of the cuboctahedron (i.e., (4/2)(12) = 24);
   $O^i_{1/3}$ = an OM_3X oxygen bonded to three of the early transition metal atoms. The four oxygen atoms of this type surround the center of the cuboctahedron at the vertices of a tetrahedron. The heteroatom X is located in the center of the cuboctahedron with tetrahedral coordination to these oxygen atoms.

The other type I structures considered by Pope[6] include $V_{10}O_{28}^{6-}$, formed by edge sharing of two V_6 macrooctahedra, and $X_2M_{18}O_{62}^{6-}$ (the so-called "Dawson structure"), formed by fusion of two M_{12} macrocuboctahedra.

These structures containing only $(\mu_n\text{-O})_5MO$ vertices can be contrasted with the non-reducible polyoxometalate structures containing only *cis*-$(\mu_n\text{-O})_4MO_2$ vertices (type II structures in the Pope nomenclature[6]). These structures are necessarily more open, since only four of the six oxygens of the MO_6 octahedra can be bridging oxygens. The most stable polyoxometalate structure is the icosahedral Silverton structure $M^{IV}(MoO^t_2O^b_{1/2}O^i_{3/3})_{12}^{8-} = M^{IV}Mo_{12}O_{42}^{8-}$ (M = Ce, Th, U), in which the central metal atom forms an MO_{12} icosahedron with the interior oxygen atoms. The central metal atom is

12-coordinate and therefore is a large tetravalent lanthanide or actinide with accessible *f* orbitals. The oxygen atoms in the Silverton structure are of the following types:

1. $O^t{}_2$ = two terminal oxygens per metal atom;
2. $O^b{}_{1/2}$= a total of six oxygen atoms ((1/2)(12) = 6) in $MMo_{12}O_{42}{}^8$, which are located at the vertices of a large octahedron;
3. $O^i{}_{3/3}$ = an $OMMo_3$ μ_4-oxygen bonded to three of the molybdenum atoms and to the central metal atom M, leading to icosahedral coordination of the latter.

An icosahedron can be decomposed into five equivalent octahedra by partitioning the 30 edges of the icosahedron into five equivalent sets of six edges each so that the midpoints of the edges in each set form a regular octahedron (Figure 10–4).[10] The vertices of the $O^b{}_6$ large octahedron in $MMo_{12}O_{42}{}^{8-}$ are located above the midpoints of the six edges in one of these octahedral sets of six icosahedron edges (Figure 10–4).

The Hückel approach (Section 4.1) can be used to treat the type I polyoxometalates as aromatic systems, in which the d_{xy} orbitals on each of the transition metal vertices overlap to form molecular orbitals. Since these d_{xy} orbitals (Figure 10–2) have two nodes, the resulting aromaticity can be called *binodal aromaticity*. Such binodal aromaticity is much weaker than anodal or uninodal orbital aromaticity (Section 4.3), since the metal atom vertices furnishing the orbitals participating in the delocalization are much further apart, being separated by M–O–M bridges rather than direct M–M bonds.

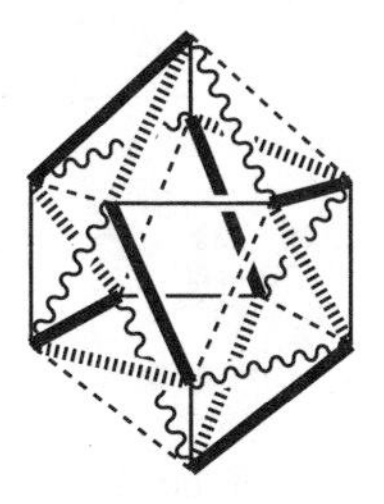

Figure 10–4: A regular icosahedron with its 30 edges partitioned into five sets of six edges each so that the midpoints of the edges in each set form a regular octahedron. The five sets of six edges each are indicated by the following types of lines:

——— ▬▬ ||||||| ∿∿∿ - - - -

The topology of the overlap of the d_{xy} orbitals can be described by a graph *G* whose vertices and edges correspond to the vertices and edges, respectively, of the macropolyhedron (octahedron or cuboctahedron). The following

[10]R. B. King and D. H. Rouvray, *Theor. Chim. Acta*, **69**, 1, 1986.

equation (≡ Equation 4–12 in Chapter 4) can then be applied to these binodal aromatic systems to relate the spectrum of G to the energy parameters of the corresponding bonding and antibonding molecular orbitals:

$$E_k = \frac{\alpha + x_k\beta}{1 + x_k S} \qquad (10\text{–}4)$$

The weakness of the binodal orbital aromaticity in type I polyoxometalates translates into a low β parameter in Equation 10–4.

Figure 10–5 illustrates the spectra of the octahedron and the cuboctahedron, which are the basic building blocks of the delocalized polyoxometalates $M_6O_{19}^{n-}$ and $XM_{12}O_{40}^{n}$, respectively. The octahedron is thus seen to have the eigenvalues +4, 0, and –2, with degeneracies 1, 3, 3, and 5, respectively. The most positive eigenvalue or *principal eigenvalue* of +4 for both polyhedra arises from the fact that each polyhedron corresponds to a regular graph of valence 4.[11] This highly positive principal eigenvalue corresponds to a highly bonding molecular orbital, which can accommodate the first two electrons upon reduction of the initially d^0 polyoxometalates of the types $M_6O_{19}^{n-}$ and $XM_{12}O_{40}^{n-}$. The reported diamagnetism[12,13] of the two-electron reduction products of the $PW_{12}O_{40}^{3-}$, $SiW_{12}O_{40}^{4-}$, and $[(H_2)W_{12}O_{40}]^{6-}$ anions is in accord with the two electrons being paired in this lowest lying molecular orbital. Thus, the overlap of the otherwise nonbonding d_{xy} orbitals in the $M_6O_{19}^{n-}$ and $XM_{12}O_{40}^{n-}$ d^0 early transition metal polyoxometalates creates a low-lying bonding molecular orbital which can accommodate two electrons, thereby facilitating reduction of polyoxometalates of these types.

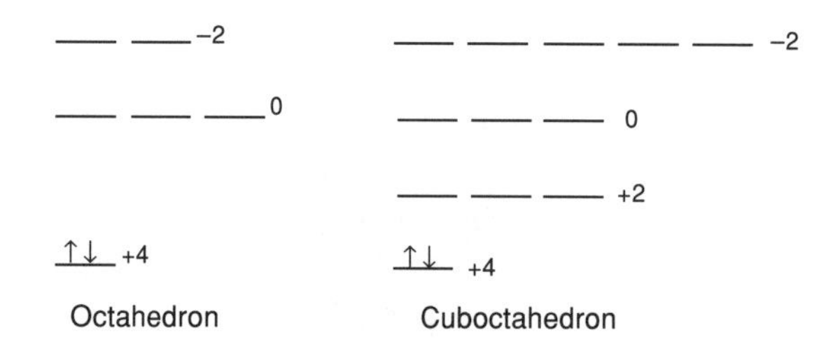

FIGURE 10–5. The spectra of the octahedron and cuboctahedron, showing the electron pairing upon two-electron reduction of macrooctahedral and macrocuboctahedral polyoxometalates.

The spectrum of the cuboctahedron (Figure 10–5), corresponding to the topology of the $XM_{12}O_{40}^{n-}$ derivatives, has not only the single +4 eigenvalue but also the triply degenerate +2 eigenvalue, corresponding to three additional bonding orbitals which can accommodate an additional six

[11] N. L. Biggs, *Algebraic Graph Theory*, Cambridge University Press, London, 1974, 14.
[12] R. A. Prados and M. T. Pope, *Inorg. Chem.*, **15**, 2547, 1976.
[13] G. M. Varga, Jr., E. Papaconstatinou, and M. T. Pope, *Inorg. Chem.*, **9**, 662, 1970.

electrons. For this reason, eight-electron reduction of the $XM_{12}O_{40}{}^{n-}$ d^0 early transition-metal derivatives might be expected to be favorable, since eight electrons are required to fill the bonding orbitals of the cuboctahedron, i.e., the four bonding orbitals corresponding to the positive eigenvalues +4 and +2. However, experimental evidence indicates that when six electrons are added to a sufficiently stable $XW_{12}O_{40}$ derivative, rearrangement occurs to a more localized $XW_9{}^{VI}W_3{}^{IV}O_{40}{}^{n-}$ structure, in which the three W^{IV} atoms form a bonded triangle[14] with W–W = 2.50 Å similar to the W–W of 2.51 Å in the tungsten(IV) complex $[W_3O_4F_9]^{5-}$. This bonded W_3 triangle corresponds to one of the triangular faces of the W_{12} macrocuboctahedron in $XW_{12}O_{40}{}^{n-}$. This rearrangement of the $XW_{12}O_{40}{}^{n-}$ derivatives to a more localized structure upon six-electron reduction is an indication of the weakness of the binodal orbital aromaticity in these polyoxometalates corresponding to a low value of β in Equation 10–4. Thus, a configuration with three W–W localized two-center two-electron σ bonds is more stable than a delocalized configuration with six electrons in the bonding molecular orbitals generated by binodal orbital overlap. Thus, if β_σ is defined by the equation

$$\beta_\sigma = (\Delta E_{\text{bonding}} - \Delta E_{\text{antibonding}})/2 \qquad (10\text{–}5)$$

for a W–W σ bond, and β_d is the energy unit in Equation 10–4 for overlap of the d_{xy} orbitals on the 12 tungsten atoms (Figure 10–2), then $\beta_\sigma >> \beta_d$.

The concept of binodal orbital aromaticity in reduced early transition metal polyoxometalates may be related to their classification as mixed valence compounds. Robin and Day[15] divide mixed valence compounds into the following three classes:

1. Class I: fully localized chemical bonding corresponding to an insulator in an infinite system;
2. Class II: partially delocalized chemical bonding corresponding to a semiconductor in an infinite system;
3. Class III: completely delocalized corresponding to a metal in an infinite system.

ESR studies on the one-electron reduced polyoxometalates $M_6O_{19}{}^{n-}$ and $XM_{12}O_{40}{}^{n-}$ suggest Class II mixed valence species.[16,17] Although such species are delocalized at accessible temperatures, they behave as localized systems at sufficiently low temperatures; this behavior is similar to that of semiconductors. This behavior is in accord with the much smaller overlap

---

[14] Y. Jeannin, J. P. Launay, and M. A. S. Sedjadi, *Inorg. Chem.*, **19**, 2933, 1980.

[15] M. B. Robin and P. Day, *Adv. Inorg. Chem. Radiochem.*, **10**, 247, 1967.

[16] J. P. Lennay, M. Fournier, C. Sanchez, J. Livage, and M. T. Pope, *Inorg. Nucl. Chem. Lett.*, **16**, 257, 1980.

[17] J. N. Barrows and M. T. Pope, *Adv. Chem. Ser.*, **226**, 403, 1990.

(i.e.., lower β in Equation 10–4) of the metal d_{xy} orbitals associated with binodal orbital aromaticity, as compared with the boron *sp* hybrid anodal internal orbitals in the deltahedral boranes $B_nH_n^{2-}$ or the carbon uninodal *p* orbitals in benzene (Section 4.3).

10.3 SUPERCONDUCTING COPPER OXIDES

Section 9.2 discusses superconductors having direct metal-metal (M–M) bonding. However, the highest T_c superconductors, namely the copper oxides, do not have direct M–M bonding but instead have indirect M–O–M bonding similar to that discussed in Section 10.1. In fact the high T_c copper oxide superconductors are excellent examples of infinite solid state metal oxide structures with relatively strong indirect metal-metal interactions through M–O–M bonding.

The well-characterized high T_c copper oxide superconductors include the following four types which can be classified by the geometries of their conducting skeletons in which electron transport takes place:

1. The 40 K superconductors $La_{2-x}M_xCuO_{4-y}$, in which the conducting skeleton consists of a single Cu–O plane;[18,19,20]
2. The 90 K superconductors $YBa_2Cu_3O_{7-y}$, in which the conducting skeleton consists of two Cu–O planes braced by a Cu–O chain;[20,21,22]
3. The homologous series of bismuth copper oxide superconductors $Bi_2Sr_2Ca_{n-1}Cu_nO_{5+2n-1-x}$ (n = 1, 2, 3), consisting of a layer sequence $BiSrCu(CaCu)_{n-1}SrBi$ and exhibiting T_c's as high as 115 K for materials isolated in the pure state;[23,24]
4. The homologous series of thallium copper oxide superconductors $Tl_2Ba_2Ca_{n-1}Cu_nO_{5+2n-1+x}$ (n = 1, 2, 3), consisting of a layer sequence

---

[18] J. G. Bednorz and K. A. Müller, *Z. Phys. B*, **64**, 189, 1986.

[19] H. H. Wang, U. Geiser, R. J. Thorn, K. D. Carlson, M. A. Beno, M. R. Monaghan, T. J. Allen, R. B. Prokasch, D. L. Stupka, W. K. Kwok, G. W. Crabtree, and J. M. Williams, *Inorg. Chem.*, **26**, 1190, 1987.

[20] J. M. Williams, M. A. Beno, K. D. Carlson, U. Geiser, H. C. Ivy Kao, A. M. Kini, L. C. Porter, A. J. Schultz, R. J. Thorn, H. H. Wang, M.-H. Whangbo, and M. Evain, *Acc. Chem. Res.*, **21**, 1, 1988.

[21] M. K. Wu, J. R. Ashburn, C. J. Torng, P. H. Hor, R. L. Meng, L. Gao, Z. J. Huang, Y. Q. Wang, and C. W. Chu, *Phys. Rev. Lett.*, **58**, 908, 1987.

[22] M.-H. Whangbo, M. Evain, M. A. Beno, and J. M. Williams, *Inorg. Chem.*, **26**, 1831 1987.

[23] R. M. Hazen, C. T. Prewitt, R. J. Angel, N. L. Ross, L. W. Finger, C. G. Hadidacos, D. R. Veblen, P. J. Heaney, P. H. Hor, R. L. Meng, Y. Y. Sun, Y. Q. Wang, Y. Y. Xue, Z. J. Huang, L. Gao, J. Bechtold, and C. W. Chu, *Phys. Rev. Lett.*, **60**, 1174 , 1988.

[24] M. A. Subramanian, C. C. Torardi, J. C. Calabrese, J. Gopalakrishnan, K. J. Morrissey, T. R. Askew, R. B. Flippen, U. Chowdhry, and A. W. Sleight, *Science*, **239**, 1015, 1988.

TlBaCu(CaCu)$_{n-1}$BaTl and exhibiting T_c's as high as 122 K for materials isolated in the pure state.[25]

All of these materials appear to consist of layers of two-dimensional Cu–O conducting skeletons separated by layers of the positive counterions, which can be regarded essentially as insulators. Resistivity[26] and critical magnetic field[27] measurements provide evidence for the two-dimensional nature of the conducting skeletons in these materials. Increasing the rigidity of the two-dimensional conducting skeleton by coupling the Cu–O layers, either by bracing with a Cu–O chain in $YBa_2Cu_3O_{7-y}$ or by close proximity in the higher members of the homologous series $M_2^{III}M_2^{II}Ca_{n-1}Cu_nO_{5+2n-1+x}$ (M^{III} = Bi, M^{II} = Sr or M^{III} = Tl, M^{II} = Ba) leads to increases in T_c. The presence of bismuth or thallium layers appears to lead to somewhat higher T_c's. This observation, coupled with computations of the electronic band structure of $Bi_2Sr_2CaCu_2O_{8+x}$[28,29] and the low resistivity of Tl_2O_3, suggests that electron transport may occur in the Bi–O or Tl–O layers as well as the Cu–O layers in the $M_2^{III}M_2^{II}Ca_{n-1}Cu_nO_{5+2n-1+x}$ materials. Destruction of the two-dimensional structure of copper oxide derivatives by gaps in the Cu–O planes or by Cu–O chains in the third dimension leads to mixed copper oxides which are metallic but not superconducting,[30] such as $La_2SrCu_2O_6$, $La_4BaCu_5O_{13}$, and $La_5SrCu_6O_{15}$.

The two-dimensional conducting skeletons in these copper oxide superconductors may be regarded as porously delocalized and otherwise analogous to those in the Chevrel phases, lanthanide rhodium borides, A-15 alloys, and other superconductors having direct M–M rather than indirect M–O–M bonding by considering the following points:

1. The conducting skeleton is constructed from Cu–O–Cu bonds rather than direct Cu–Cu bonds. The much higher ionic character and thus much lower polarizability and higher rigidity of metal-oxygen bonds relative to metal-metal bonds can be related to the persistence of superconductivity in copper oxides to much higher temperatures than that in metal clusters.

---

[25]R. M. Hazen, L. W. Finger, R. J. Angel, C. T. Prewitt, N. L. Ross, C. G. Hadidacos, P. J. Heaney, D. R. Veblen, Z. Z. Sheng, A. El Ali, and A. M. Hermann, *Phys. Rev. Lett.*, **60**, 1657, 1988.

[26]S. W. Cheong, Z. Fisk, R. S. Kwok, J. P. Remeika, J. D. Thomson, and C. Gruner, *Phys. Rev. B*, **37**, 5916, 1988.

[27]J. S. Moodera, R. Meservey, J. E. Tkaczyk, C. X. Hao, G. A. Gibson, and P. M. Tedrow, *Phys. Rev. B*, **37**, 619, 1988.

[28]M. S. Hybertsen and L. F. Mattheiss, *Phys. Rev. Lett.*, **60**, 1661, 1988.

[29]H. Krakauer and W. E. Pickett, *Phys. Rev. Lett.*, **160**, 1665, 1988.

[30]J. B. Torrance, Y. Tokura, A. Nazzal, and S. S. P. Parkin, *Phys. Rev. Lett.*, **60**, 542, 1988.

2. The required metal-metal interactions are antiferromagnetic interactions between the single unpaired electrons in the d_{xy} orbitals of two d^9 Cu(II) ions, separated by an oxygen bridge similar to antiferromagnetic Cu(II)–Cu(II) interactions in discrete binuclear complexes (Section 10–1).[4,31] This general idea has been presented by Anderson and co-workers as the so-called resonating valence bond model.[32,33]
3. The positive counterions in the copper oxide superconductors control the negative charge on the Cu–O skeleton and hence the oxidation states of the copper atoms. Positive counterions, which are "hard" in the Pearson sense[34] in preferring to bind to nonpolarizable bases, such as the lanthanides and alkaline earths, do not contribute to the conductivity. However, layers of the positive counterions bismuth and thallium, which are "soft" in the Pearson sense in preferring to bind to polarizable bases, may contribute to the conductivity as noted above.
4. Partial oxidation of some of the Cu(II) to Cu(III) in the copper oxide superconductors listed above generates holes in the conduction band required for conductivity. This is in accord with Hall effect measurements[35] which show positive Hall coefficients indicating that holes rather than electrons are the current carriers. In addition, more recent work[36,37] has resulted in the discovery of 28 K superconductors having the general formula $Ln_{2-x}Ce_xCuO_{4-y}$ (Ln = Pr, Nd, Sm). These superconductors have negative Hall coefficients, indicating that mobile electrons are the current carriers. These mobile electrons arise by partial reduction of some of the Cu(II) to Cu(I).

---

[31] C. J. Cairns and D. H. Busch, *Coord. Chem. Rev.*, **69**, 1, 1986.

[32] P. W. Anderson, *Science*, **235**, 1196, 1987.

[33] P. W. Anderson, G. Baskaran, Z. Zou, and T. Hsu, *Phys. Rev. Lett.*, **58**, 2790, 1987.

[34] R. G. Pearson, *J. Am. Chem. Soc.*, **85**, 3533, 1963.

[35] N. P. Ong, Z. Z. Wang, J. Clayhold, J. M. Tarascon, L. H. Greene, and W. R. McKinnon, *Phys. Rev. B*, **35**, 8807, 1987.

[36] Y. Tokura, H. Takagi, and S. Uchida, *Nature*, **337**, 345, 1989.

[37] H. Sawa, S. Suzuki, M. Watanabe, J. Akimitsu, H. Matsubara, H. Watabe, S. Uchida, K. Kokusho, H. Asano, F. Izumi, and E. Takayama-Muromachi, *Nature*, **337**, 347, 1989

Chapter 11

THE ICOSAHEDRON IN INORGANIC CHEMISTRY: BORON ALLOTROPES, ICOSAHEDRAL QUASICRYSTALS, AND CARBON CAGES

11.1 SYMMETRY OF THE ICOSAHEDRON

Among the polyhedra which arise in chemical contexts, the regular icosahedron (Figure 4–5) is of particular interest in terms of both its symmetry and topology. The symmetry point group (Section 2.2),[1] I_h, of the icosahedron, which has 120 operations, is the largest non-trivial point group; its rotation subgroup, I, which has 60 operations, is isomorphic to the *alternating group* A_5, of even permutations of five objects. Note that the icosahedral point groups I and I_h are the only symmetry point groups having multiple three-fold (C_3) and five-fold (C_5) rotation axes.

Let us now consider in greater detail the properties of the transitive permutation groups (Section 2–5) on small numbers of objects. In this context, the *symmetric* group P_n is a group consisting of all possible $n!$ permutations of n objects, whereas the *alternating* group A_n is a group consisting of all possible $n!/2$ *even* permutations of n objects. The alternating group A_3 and the symmetric group P_3 on three objects are isomorphic with the point groups C_3 and D_3, respectively. The latter two point groups correspond to the symmetry of the two-dimensional simplex,[2] i.e., to that of a planar equilateral triangle. Similarly, the alternating group A_4 and symmetric group P_4 on four objects are, respectively, isomorphic with the point groups T and T_d, which correspond to the symmetry of the three-dimensional simplex, i. e., to that of the regular tetrahedron. The alternating group on five objects, A_5, has $5!/2 = 60$ operations, like the icosahedral pure rotation group I. Similarly, both the symmetric group P_5 and the full icosahedral point group, I_h, have $5! = 120$ operations. Examination of the conjugacy class structure (Section 2.3)[3,4] of the permutation groups A_5 and P_5, on the one hand, and of the icosahedral point groups I and I_h, on the other hand, reveals that they correspond to each other by the relationships:

$$I \cong A_5 \text{ (isomorphism)} \tag{11–1}$$

$$P_5 = A_5 \wedge P_2 \text{ (semi-direct product)} \tag{11–2}$$

---

[1] F. A. Cotton, *Chemical Applications of Group Theory*, John Wiley & Sons, New York, 1971.

[2] B. Grünbaum, *Convex Polytopes*, Interscience, New York, 1967.

[3] F. J. Budden, *The Fascination of Groups*, Cambridge University Press, London, 1972.

[4] T. Janssen, *Crystallographic Groups*, North Holland, Amsterdam, 1973, 7.

$$I_h = I \times C_2 \text{ (direct product)} \tag{11–3}$$

In this context the direct product G (compare Section 2.4) of two groups A and B (i.e., $G = A \times B$) can be defined as:

1. For any $a \in A$ and any $b \in B$ the automorphism $\phi(b)$ of A is the identity, thus

$$\phi(b)a = a \tag{11–4}$$

2. There is an isomorphism between G and the group of pairs (a,b) with $a \in A$ and $b \in B$ which satisfies the multiplication law

$$(a_1,b_1)(a_2,b_2) = (a_1a_2,b_1b_2) \tag{11–5}$$

Similarly a group G is a *semi-direct product* of two groups A and B (i.e., $G = A \wedge B$) when:

1. For any $a \in A$ and any $b \in B$ there is an automorphism $\phi(b)$ of A such that

$$\phi(b_1)[\phi(b_2)a] = \phi(b_1b_2)a \tag{11–6}$$

2. There is an isomorphism between G and the group of pairs (a,b) with $a \in A$ and $b \in B$, which satisfies the multiplication law

$$(a_1,b_1)(a_2,b_2) = (a_1[\phi(b_1)a_2],b_1b_2) \tag{11–7}$$

A direct product is thus a special case of a semi-direct product.

Let us now consider the P_n symmetric permutation groups having $n!$ operations. In these groups, as well as in any other permutation group, permutations having different cycle structures necessarily belong to different conjugacy classes. However, in the symmetric groups P_n, a common partition of n will guarantee that the elements do in fact belong to the same class. Thus, the conjugacy classes of the permutation groups A_5 and P_5 are indicated in terms of their cycle indices (Section 2.5)[5,6] $Z(G)$ in the following way:

$$60\ Z(A_5) = x_1^5 + 20x_1^2x_3 + 15x_1x_2^2 + 24x_5 \tag{11–8}$$

$$120\ Z(P_5) = x_1^5 + 10x_1^3x_2 + 20\ x_1^2x_3 + 15x_1x_2^2 + 30\ x_1x_4 + 20\ x_2x_3 + 24x_5 \tag{11–9}$$

---

[5]G. Pólya, *Acta Math.*, 68 145, 1937.

[6]N. G. Debruin in *Applied Combinatorial Mathematics*, E. F. Bechenback, Ed., John Wiley Sons, New York, 1964.

Furthermore, the conjugacy classes of the icosahedral point groups I and I_h are indicated from their character tables to be the following:

$$I = \{E,\ 12C_5,\ 12\ C_5^2,\ 20\ C_3,\ 15C_2\} \tag{11–10}$$

$$I_h = \{\mathrm{E},\ 2C_5,\ 12\ C_5^2,\ 20\ C_3,\ 15C_2,\ i,\ 12S_{10}^3,\ 12S_{10},\ 20S_6,\ 15\sigma\} \tag{11–11}$$

Comparison of Equations 11–8 and 11–9 with Equations 11–10 and 11–11, respectively, by using the relationships in Equations 11–1, 11–2, and 11–3, leads to the following observations:

1. The class of A_5 represented by the cycle index term $24x_5$ corresponds to the two classes $12C_5$ and $12C_5^2$ of I taken together.
2. In the point group I_h, each class of improper rotations (i, S_{10}, S_{10}^3, S_6, and σ) corresponds to a class of proper rotations of the same size, namely to E, C_5, C_5^2, C_3, and C_2, respectively, whereas the classes of P_5 are not partitioned analogously.

The relationship between the icosahedral point groups I_h and I outlined here and the permutation groups A_5 and P_5 is discussed in more detail elsewhere.[7]

11.2 BORON ICOSAHEDRA IN ELEMENTAL BORON AND METAL BORIDES

Section 4.3 discusses globally delocalized icosahedral boranes such as $B_{12}H_{12}^{2-}$. Boron icosahedra can also be used to construct infinite solid state structures, including the allotropes of elemental boron as well as some boron-rich metal borides. Understanding the structure and bonding in infinite solid state structures based on icosahedral B_{12} building blocks, is also relevant to understanding the icosahedral quasicrystals discussed in the next section, which are constructed from aluminum icosahedra. Note that an icosahedral B_{12} building block, in which each of the twelve vertices contributes a single electron for an external two-electron two-center bond to an external group, requires a total of 38 electrons distributed as follows:

Electrons for each of the 12 external bonds:	12 electrons
Electrons for the 12-center core bond of the B_{12} icosahedron:	2 electrons
Electrons for the surface bonding:	24 electrons
Total electrons required:	*38 electrons*

[7] R. B. King and D. H. Rouvray, *Theor. Chim. Acta*, 69 1, 1986.

Since 12 boron atoms have a total of (12)(3) = 36 valence electrons, such an icosahedral B_{12} unit is stable as the formal dianion B_{12}^{2-}.

Elemental boron exists in a number of allotropic forms, of which four (two rhombohedral forms and two tetragonal forms) are well established.[8,9,10] The structures of all of these allotropic forms of boron are based on various ways of joining B_{12} icosahedra using the external orbitals on each boron atom. The structures of the two rhombohedral forms of elemental boron are of interest in illustrating what can happen when icosahedra are packed into an infinite three-dimensional lattice. Note that in rhombohedral structures, the local symmetry of an icosahedron is reduced from I_h to D_{3h} because of the loss of the five-fold rotation axes. The twelve vertices of an icosahedron, which are all equivalent under I_h local symmetry, are split under D_{3h} local symmetry into two non-equivalent sets of six vertices each (Figure 11–1). The six rhombohedral vertices (labeled R in Figure 11–1) define the directions of the rhombohedral axes. The six equatorial vertices (labeled E in Figure 11–1) lie in a staggered belt around the equator of the icosahedron. The six rhombohedral and six equatorial vertices form prolate (elongated) and oblate (flattened) trigonal antiprisms, respectively.

FIGURE 11–1. The six rhombohedral (labeled R) and equatorial (labeled E) vertices of an icosahedron.

In the simple (α) rhombohedral allotrope of boron, all boron atoms are part of discrete icosahedra. In a given B_{12} icosahedron, the external orbitals of the rhombohedral boron atoms (R in Figure 11–1) form two-center bonds with rhombohedral boron atoms of an adjacent B_{12} icosahedron, and the external orbitals of the equatorial boron atoms (E in Figure 11–1) form three-center bonds with equatorial borons of two adjacent B_{12} icosahedra. The available (12)(3) = 36 electrons from an individual B_{12} icosahedron in α-rhombohedral boron are fully used as follows:

[8]J. L. Hoard and R. E. Hughes in *The Chemistry of Boron and its Compounds*, Ed. E. L. Muetterties, John Wiley & Sons, New York, 1967, 25.

[9]D. Emin, T. Aselage, C. L. Beckel, I. A. Howard, C. Wood, Eds., *Boron-Rich Solids*, American Institute of Physics Conference Proceedings 140, American Institute of Physics, New York, 1986.

[10]W. N. Lipscomb and D. Britton, *J. Chem. Phys.*, 33, 275, 1960.

Skeletal bonding	
12-center core bond:	2 electrons
12 2-center surface bonds: (12)(2) =	24 electrons
External bonding	
(a) Rhombohedral borons:	
1/2 of six 2-center bonds: (6/2)(2) =	6 electrons
(b) Equatorial borons:	
1/3 of six 3-center bonds: (6/3)(2) =	4 electrons
Total electrons required:	*36 electrons*

α-Rhombohedral boron thus has a closed-shell electronic configuration.

The structure of the complicated (β) rhombohedral allotrope of boron avoids the three-center intericosahedral bonding of α-rhombohedral boron but is considerably more complicated.[11] The structure of β-rhombohedral boron may be described as a rhombohedral packing of B_{84} polyhedral networks known as Samson complexes (Figure 11–2),[12] linked by B_{10} polyhedra and an interstitial boron atom so that the fundamental structural unit is $B_{84}(B_{10})_{6/3}B = B_{105}$. The idealized isolated B_{84} Samson complexes have I_h local symmetry, which is distorted to D_{3h} in the rhombohedral local environment of the lattice. Within the B_{84} Samson complex, the external orbital of each of the twelve boron atoms of a central B_{12} icosahedron forms a two-center bond with the external orbital of an apical boron atom of a B_6 pentagonal pyramid (i.e., a half icosahedron), leading to the $B_{12}(B_6)_{12} = B_{84}$ stoichiometry of this B_{84} Samson complex. The external surface of this B_{84} Samson complex (Figure 11–2) is a B_{60} truncated icosahedron identical to the C_{60} truncated icosahedron of fullerene.[13] The B_6 pentagonal pyramids in the rhombohedral positions (see R of Figure 11–1) of the central B_{12} icosahedron of the B_{84} Samson complex overlap with analogous B_6 pentagonal pyramids of adjacent B_{84} Samson complexes to form six new B_{12} icosahedral cavities. The B_6 pentagonal pyramids in the equatorial positions (see E of Figure 11–1) of the central B_{12} icosahedron of the B_{84} Samson complexes each overlap with the corresponding equatorial B_6 pentagonal pyramids of two adjacent B_{84} Samson complexes by means of an additional B_{10} unit to form new polyhedra of 28 boron atoms (Figure 11–3). These B_{28} units have local C_{3v} symmetry and are constructed by fusion of three icosahedra, so that in each icosahedron, one vertex (vertex A in Figure 11–3) is shared by all three icosahedra and four vertices (B and D in Figure 11–3) are each shared by two of the icosahedra so that $3(B_7B_{4/2}B_{1/3}) = B_{28}$.

[11] R. B. King, *Inorg. Chim. Acta*, **181**, 217, 1991.

[12] L. Pauling, *Phys. Rev. Lett.*, 58, 365, 1987.

[13] H. W. Kroto, A. W. Allof, and S. P. Balin, *Chem. Rev.*, 91 , 1213, 1991.

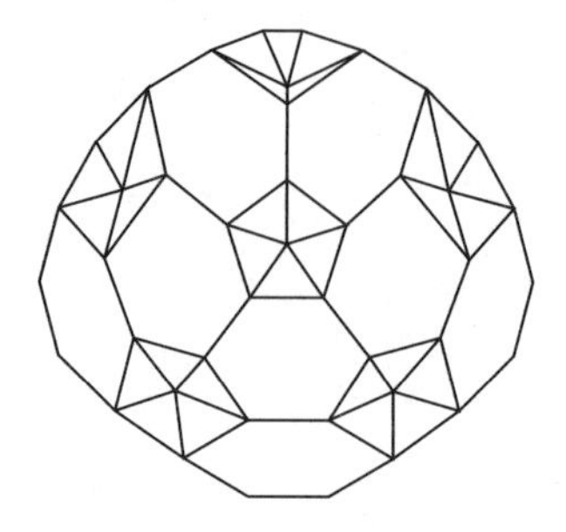

FIGURE 11–2. A view of the surface of a Samson complex showing six of the twelve pentagonal pyramid cavities.

In order to consider an electron counting scheme for β-rhombohedral boron, it is first necessary to consider the chemical bonding topology of the idealized B_{28} polyhedron (Figure 11–3) formed by the fusion of three globally delocalized boron icosahedra. The 28 boron atoms furnish a total of (28)(4) = 112 valence orbitals, of which 24 orbitals (one on each boron atom except for the four boron atoms labelled A and B in Figure 11–3) are required for external bonding, leaving 112 – 24 = 88 atomic orbitals for the skeletal (internal) bonding. A 12-center core bond in each of the icosahedral cavities of the B_{28} polyhedron requires (3)(12) = 36 atomic orbitals, leaving 88 – 36 = 52 atomic orbitals for pairwise surface bonding, corresponding to 26 surface bonds. Thus, a closed shell electronic configuration for the B_{28} polyhedron with one electron from the boron vertex in each external orbital is B_{28}^{2+}, which requires 82 electrons as follows:

24 external two-center bonds: (24/2)(2)	24 electrons
3 12-center core bonds: (3)(2) =	6 electrons
26 surface bonds: (26)(2) =	52 electrons
Total electrons required:	82 electrons

Three boron positions (D in Figure 11–3) in the B_{28} polyhedron are only partially occupied (~$^2/_3$) because of the availability of only four valence orbitals on the interstitial boron for chemical bonding. This corresponds approximately to removing one of these boron atoms from each B_{28} polyhedron. Removal of this boron atom from the B_{28} polyhedron to give a B_{27} polyhedron removes three electrons and four orbitals. Loss of these four orbitals has the following three effects

1. One external bond is eliminated, reducing the total required number of electrons by one;
2. One core bond is reduced from a 12-center bond to an 11-center bond with no effect on the required number of electrons;

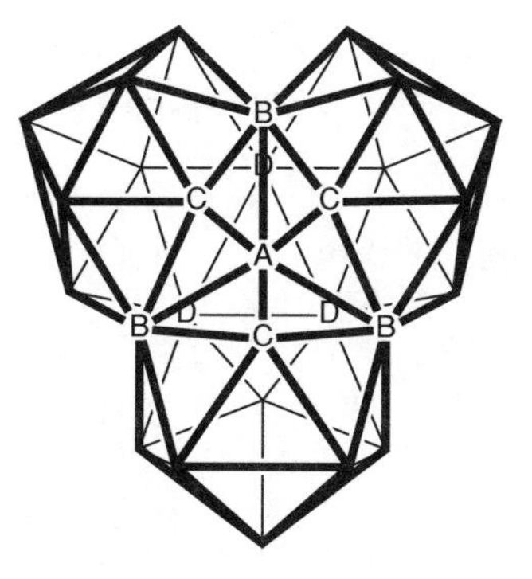

FIGURE 11–3. The B_{28} polyhedron, linking equatorial positions of three Samson complexes in the β-rhombohedral boron structure.

3. One surface bond is eliminated, reducing the required number of electrons by two. Thus, the removal of one D vertex in the B_{28} polyhedron removes three electrons but also the need for three electrons (3 = 1 + 0 + 2 from (1), (2), and (3) above, respectively), so that the net charge on the species with the closed shell electronic configuration is not affected.

Additional complicating features of the structure of β-rhombohedral boron are the partial occupancy of some boron sites and the presence of interstitial boron atoms as follows[14]:

1. Three of the boron vertex sites of the B_{10} unit linking three B_6 pentagonal pyramids to form the B_{28} polyhedron, namely the vertices marked "D" in Figure 11–3, are only partially occupied (73.4%).
2. There is an interstitial boron atom (designated as B(15) in the structural papers) within bonding distance of six of the above partially occupied boron vertex sites, corresponding approximately to an isolated tetracoordinate boron atom (i.e., (0.734)(6) = 4.4).
3. There is also a partially occupied (24.8%) interstitial site in the B_{84} Samson complex.

The boron atom in the fully occupied interstitial site (B(15) in the structural papers) bonded to four boron atoms of the B_{27} polyhedron (D in Figure 11–3) has a closed shell configuration B^- (compare BH_4^- or $B(C_6H_5)_4^-$). The ~25% occupancy of the other six interstitial sites in each B_{84} Samson complex (B(16) in the structural papers) provides another $B_{1.5}$ for the fundamental structural unit. This additional $B_{1.5}$ provides an extra (1.5)(3) = 4.5 electrons without adding any new bonding orbitals, since the atomic orbitals of these

[14]B. Callmer, *Acta Cryst.*, 33B, 1951, 1977.

latter interstitial boron atoms merely increase some two-center surface bonds to three-center bonds.

All of these considerations lead to a very complicated model for the chemical bonding topology of β-rhombohedral boron. The electron counting in its fundamental $B_{104.5}$ structural unit is summarized in Table 11–1. The net charge of –0.5 for a 313.5 (= (3)(104.5)) valence electron structural unit can be assumed to be zero within the experimental error of partial occupancies, etc., indicating that β-rhombohedral boron, like the much simpler α-rhombohedral boron, has a closed shell electronic configuration. Note that in the lattices of both rhombohedral forms of boron, the six rhombohedral (R in Figure 11–1) and six equatorial (E in Figure 11–1) borons of a central B_{12} icosahedron are linked to two and three other such B_{12} icosahedra, respectively, using simple two-center and three-center chemical bonds, respectively, for α-rhombohedral boron but using B_{12} icosahedra and B_{28} polyhedra (Figure 11–3), respectively, for β-rhombohedral boron.

TABLE 11–1
Boron and Electron Counting in β-Rhombohedral Boron

	Boron Atoms	Net Charge
Central B_{12} icosahedron	12	–2
6/2 Rhombohedrally located peripheral B_{12} icosahedra:		
(6/2)(12) =	36	
(6/2)(–2) =		–6
6/3 Equatorially located peripheral B_{27} polyhedra:		
(6/3)(27) =	54	
(6/3)(+2) =		+4
1 B(15) interstitial boron atom:	1	–1
(0.25)(6) = 1.5 B(16) interstitial boron atoms:		
(1.5)(1) =	1.5	
(1.5)(+3) =		+4.5
Total boron atoms and overall net charge:	*104.5*	*–0.5*

Boron icosahedra are also found in boron-rich borides of the most electropositive metals, such as lithium, sodium, magnesium, and aluminum.[15] The structures of such borides are related to the alkali metal gallides discussed in Section 8.2. In the structures of these boron rich-borides, the electropositive metals can be regarded as forming cations, e.g., Li^+, Na^+, Mg^{2+}, etc., and the boron subnetwork then has a corresponding negative charge. An important structural unit in such borides is B_{14}^{4-}, which may be written more precisely as $(B_{12}^{2-})(B^-)_2$. Thus, consider the magnesium boride Mg_2B_{14} = $(Mg^{2+})_2(B_{12}^{2-})(B^-)_2$.[16] Half of the external bonds from the B_{12}

[15] R. Naslain, A. Guette, and P. Hagenmuller, *J. Less Common Metals*, 47, 1, 1976.
[16] A. Guette, M. Barret, R. Naslain, P. Hagenmuller, L. E. Tergenius, and T. Lundström, *J.*

icosahedra are direct bonds to other B_{12} icosahedra, whereas the other half of these external bonds are to the isolated boron atoms. Closely related structures are found in $LiAlB_{14}$[17] and the so-called "$MgAlB_{14}$." However, "$MgAlB_{14}$" is actually $MgAl_{2/3}B_{14}$ because of partial occupancy of the aluminum sites and thus can be formulated with the same B_{14}^{4-} unit as Mg_2B_{14}.[18] A less closely related structure is $NaB_{0.8}B_{14}$, which has two types of interstitial boron atoms as well as the same B_{12}^{2-} icosahedra.[19]

The lanthanides are also examples of electropositive metals that form boron-rich borides, having boron subnetworks constructed from B_{12} icosahedra. An example of an extremely boron-rich metal boride is YB_{66}.[20,21] The structure of YB_{66} is even more complicated than than that of β-rhombohedral boron discussed above and has not been worked out in complete detail. The unit cell of YB_{66} has approximately 24 yttrium atoms and 1584 boron atoms. The majority of the boron atoms (1248 = (8)(156)) are contained in thirteen-icosahedron units of 156 atoms each. In such a thirteen-icosahedron unit, a central B_{12} icosahedron is surrounded by 12 icosahedra, leading to a B_{156} "icosahedron of icosahedra." The remaining boron atoms are statistically distributed in channels that result from the packing of the thirteen-icosahedron units and form nonicosahedral cages, which are not readily characterized. The complexity of this structure and the uncertainty in the positions of the "interstitial" boron atoms clearly preclude any serious attempts at electron counting.

Another interesting solid state boron compound is the refractory material boron carbide which has stoichiometries B_4C (= $B_{12}C_3$)[22] to $B_{13}C_2$.[23] These compounds have an interesting structure, in which B_{12} icosahedra are linked by planar C_2B_4 six-membered rings similar to the six-membered rings in the planar graphite (Figure 11–4). The planarity of the six-membered C_2B_4 rings suggests "benzenoid-type" aromaticity. Additional carbon (in $B_{12}C_3$) or boron (in $B_{13}C_2$) atoms link the carbon atoms to form allene-like C_3 or CBC chains, respectively, in directions perpendicular to the planes of the C_2B_4 rings. The combination of the very stable "aromatic" hexagonal C_2B_4 rings fused to the likewise very stable B_{12} icosahedra can account for the observed stability and strength of the $B_{12}C_3$ and $B_{13}C_2$ structures, including the extreme hardness and high melting points of these boron carbides.

Less Common Metals, 82, 325, 1981

[17]I. Higashi, *J. Less Common Metals*, 82, 317, 1981.

[18]V. I. Matkovich and J. Economy, *Acta Cryst.*, 26B, 616, 1969.

[19]R. Naslain and J. S. Kasper, *J. Solid State Chem.*, 1, 150, 1970.

[20]S. M. Richards and J. S. Kasper, *Acta Cryst.*, 25B, 257, 1969.

[21]J. S. Kasper, *J. Less Common Metals*, 47, 17., 1976

[22]K. H. Clark and J. L. Hoard, *J. Am. Chem. Soc.*, 65, 2115, 1943.

[23]G. Will and K. H. Kossobutzki, *J. Less Common Metals*, 44, 87, 1976.

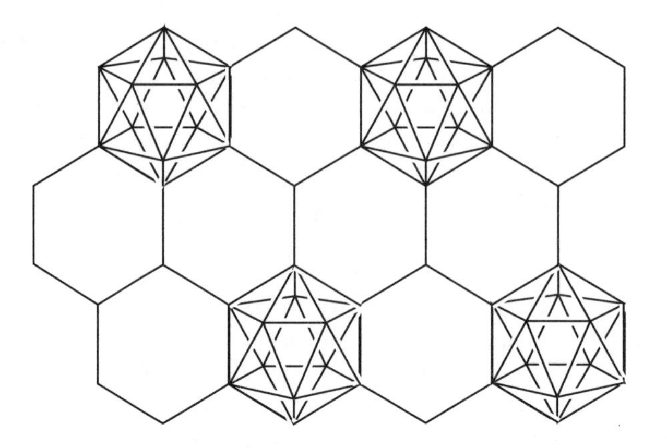

FIGURE 11–4. A hexagonal layer in $B_{12}C_3$ or $B_{13}C_2$, showing the fusion of the B_4C_2 planar hexagons to the B_{12} icosahedra.

11.3 ICOSAHEDRAL QUASICRYSTALS

One of the most exciting recent developments in materials science has been the discovery of aluminum alloys exhibiting diffraction patterns with apparently sharp spots containing five-fold symmetry axes.[24,25] This discovery raises a crystallographic dilemma, since the sharpness of the diffraction peaks suggests long-range translational order, as in periodic crystals, but five-fold axes are incompatible with such periodicity. Such materials are described as *quasicrystals*,[26,27] which are defined to have delta functions in their Fourier transforms but local point symmetries incompatible with periodic order. The structures of these materials may be viewed as three-dimensional analogues of Penrose tiling,[28,29,30,31] which is a geometric structure exhibiting five-fold symmetries and Bragg diffraction. Location of atoms in quasicrystals requires the use of six-dimensional crystallography,[32,33] in which the atoms correspond to three dimensional hypersurfaces in six-dimensional periodic lattices, rather than normal three-dimensional crystallography, in which the atoms correspond to zero-dimensional points in three-dimensional periodic lattices. Note that in both

---

[24] D. Shechtman, I. Blech, D. Gratias, and J. W. Cahn, *Phys. Rev. Lett.*, **53**, 1951, 1984.
[25] D. Shechtman and I. A. Blech, *Metall. Trans.*, **16A**, 105, 1985.
[26] D. Levine and P. J. Steinhardt, *Phys. Rev. Lett.*, **53**, 2477, 1984.
[27] D. Levine and P. J. Steinhardt, *Phys. Rev. B*, **34**, 596, 1986.
[28] N. de Bruijn, *Ned. Akad. Weten. Proc. Ser. A.*, **43**, 39, 1981.
[29] N. de Bruijn, *Ned. Akad. Weten. Proc. Ser. A.*, **43**, 53, 1981.
[30] A. L. Mackay, *Sov. Phys. Crystallogr.*, **26**, 517, 1981.
[31] A. L. Mackay, *Phys. Status Solidi A*, **114**, 609, 1982.
[32] V. Elser and C. L. Henley, *Phys;. Rev. Lett.*, **55**, 2883, 1985.
[33] P. Bak, *Phys. Rev. Lett*, **56**, 861, 1986.

the conventional three-dimensional crystallography used to determine the structures of regular crystals and the six-dimensional crystallography required to determine the structures of quasicrystals, the dimensionality of the locations of the atoms is three less than the dimensionality of the lattice.

The complications of six-dimensional crystallography prevent the chemical structures of quasicrystals from being described in ways familiar to chemists. Instead, chemical models for quasicrystal structures are more readily developed by first considering closely related true crystalline materials and then introducing appropriate perturbations destroying the periodic translational order but retaining the long-range translational and orientational order characteristic of quasicrystals.[26,27]

There are several classes of icosahedral quasicrystals with diverse compositions.[34] The following classes are the most important:

1. The i(Al–Mt) class (Mt = transition metal), including i($Al_{80}Mn_{20}$),[24] i($Al_{74}Mn_{20}Si_6$),[35,36] and i($Al_{79}Cr_{17}Ru_4$)[37];
2. The i(AlZnMg) class, including i($Al_{25}Zn_{38}Mg_{37}$),[38] i($Al_{44}Zn_{15}Cu_5Mg_{36}$),[39] and i($Al_{60}Cu_{10}Li_{30}$).[40]

All of these phases contain relatively large amounts of aluminum, suggesting the comparison of the structures of these materials with those of compounds with boron and gallium icosahedra, including the icosahedral cage boranes (Section 4.3), elemental boron and boron-rich metal borides (Section 11.2), and gallium-rich intermetallic phases of alkali metals and gallium (Section 8.2).

Consider the so-called T2 icosahedral quasicrystal i($Al_{60}Cu_{10}Li_{30}$) and related alloys containing aluminum, lithium, and copper and/or zinc, which have been studied by Audier and collaborators.[41] These alloys are significant since there are true crystalline phases having stoichiometries similar to these T2 icosahedral quasicrystals. Thus, the atom positions in the crystalline cubic R phase of approximate Al_5CuLi_3 stoichiometry are known, so that the structure of this phase can provide some insight into the structure of the closely related T2 icosahedral phase of approximate stoichiometry $Al_{0.570}Cu_{0.108}Li_{0.322}$. Of interest in the structure of R-Al_5CuLi_3 is the presence of Al_{12} icosahedra. Discrete molecules containing similar Al_{12}

---

[34] C. L. Henley, *Comments Cond. Mater. Phys.*, **13**, 59, 1987.

[35] P. Guyot and M. Audier, *Philos. Mag.*, **B52**, L15, 1985.

[36] L. A. Bendersky and M. J. Kaufman, *Philos. Mag.*, **B53**, L75, 1986.

[37] P. A. Bancel and P. A. Heiney, *Phys. Rev. B*, **33**, 7919, 1986.

[38] R. Ramachandrarao and G. V. S. Sastry, *Pramana*, **25**, L225, 1985.

[39] N. K. Mukhopadhyay, G. N. Subbanna, S. Ranganathan, and K. Chattopadhyay, *Scr. Metall.*, **20**, 525, 1986.

[40] P. Sainfort, B. Dubost, and A. Dubus, *Comp. Rend. Acad. Sci. Paris*, **301**, 689, 1985.

[41] M. Audier, C. Janot, M. de Boissieu, and B. Dubost, *Philos. Mag.*, **B60**, 437, 1989.

icosahedra are very rare, although i-$Bu_{12}Al_{12}^{2-}$, containing an Al_{12} icosahedron and isoelectronic with $B_{12}H_{12}^{2-}$, has recently been reported as a low yield product from the reduction of i-Bu_2AlCl with potassium metal.[42]

Audier and collaborators[41,43] have described a polyhedral shell structure for R-Al_5CuLi_3, using a number of polyhedra of icosahedral symmetry. This shell structure consists of the following layers:

(a) a central $(Al,Cu)_{12}$ icosahedron;
(b) an Li_{20} regular dodecahedron, with the Li positions above the faces of the central $(Al,Cu)_{12}$ icosahedron (layer a), analogous to the construction of dual polyhedra (Section 1.4);
(c) a larger $(Al,Cu)_{12}$ icosahedron, formed from the external orbitals of the central $(Al,Cu)_{12}$ icosahedron (layer a) so that its atoms lie above the twelve faces of the Li_{20} dodecahedron in another "polyhedral dual construction";
(d) an $(Al,Cu)_{60}$ truncated icosahedron (Figure 11–5), distorted from I_h symmetry to O_h symmetry so that 12 vertices are of one type (starred in Figure 11–5) and 48 vertices are of another type (not starred in Figure 11–5). The atoms of this polyhedron lie above the midpoints of the faces of the omnicapped dodecahedron formed by combining the Li_{20} dodecahedron (layer b) with the larger $(Al,Cu)_{12}$ icosahedron (layer c) in still another "polyhedral dual construction," since the truncated icosahedron is the dual of the omnicapped dodecahedron.

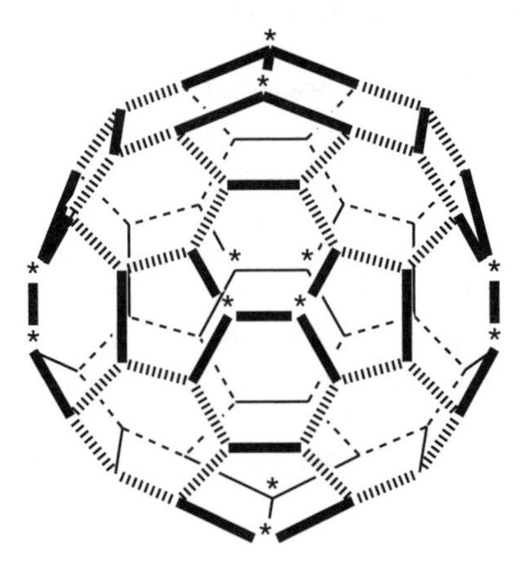

FIGURE 11–5. The two different types of vertices in the truncated icosahedron distorted from I_h to O_h symmetry, as found in the structure of R-Al_5CuLi_3. For clarity, the edges of the hexagonal faces *not* containing any starred vertices are indicated by dotted lines.

---

[42]W. Hiller, K.-W. Klinkhammer, W. Uhl, and J. Wagner, *Angew. Chem. Int. Ed.*, **30**, 179, 1991.

[43]M. Audier, J. Pannetier, M. Leblanc, C. Janot, J.-M. Lang, and B. Dubost, *Phys. Status Solidi B*, **153**, 136, 1988.

The (Al,Cu) subskeleton (layers a + c + d discussed above) in R-Al_5CuLi_3 forms a lattice containing 84-vertex Samson complexes (Figure 11–2) identical to the B_{84} Samson complex in the β-rhombohedral boron lattice discussed in the previous section. However, the packing of the $(Al,Cu)_{84}$ Samson complexes in the R-Al_5CuLi_3 lattice is totally different from the packing of the B_{84} Samson complexes in β-rhombohedral boron. The (Al,Cu) subskeleton of the R-Al_5CuLi_3 lattice thus consists of a CsCl-type cubic packing of the $(Al,Cu)_{84}$ Samson complexes, so that each atom of the peripheral $(Al,Cu)_{60}$ truncated icosahedron of the $(Al,Cu)_{84}$ Samson complex is shared with an adjacent Samson complex in one of the following two ways:

1. The six edges connecting the six pairs of starred vertices in Figure 11–5 lie in the faces of a cube and are shared with the corresponding edges of the adjacent Samson complex in the cube sharing the face containing the edge in question;
2. The eight hexagonal faces of the peripheral $(Al,Cu)_{60}$ truncated icosahedron *not* containing any starred vertices in Figure 11–5 (these eight hexagonal faces are indicated by dotted edges in Figure 11–5) are shared with the corresponding hexagonal faces of the adjacent Samson complex in the cube sharing the vertex lying at the end of the body diagonal of the original cube containing the midpoint of the hexagonal face in question.

The fundamental structural unit of the (Al,Cu) subskeleton of R-Al_5CuLi_3, thus, is $(Al,Cu)_{54}$ = $(Al,Cu)_{12}$ + $(Al,Cu)_{12}$ + $(Al,Cu)_{60/2}$, corresponding, respectively, to the layers a + c + d discussed above. In addition to the 20 lithium atoms in layer b, additional lithium atoms (layer e) are located above the twelve pentagonal faces of the peripheral truncated icosahedra which, because of the way that the truncated icosahedra are linked, simultaneously cap the pentagonal faces of two adjacent truncated icosahedra sharing an edge (sharing method 1 above). The sum of the number of atoms in layers a through e, respectively, in this model is $(Al,Cu)_{12}$ + Li_{20} + $(Al,Cu)_{12}$ + $(Al,Cu)_{60/2}$ + $Li_{12/2}$ = $(Al,Cu)_{54}Li_{26}$, corresponding to an (Al + Cu)/Li ratio of 2.08, in close agreement with 2.12 implied by the $Al_{0.564}Cu_{0.116}Li_{0.32}$ stoichiometry.

Now consider the electron counting in R-Al_5CuLi_3. In the (Al,Cu) subskeleton, the aluminum and copper atoms function as donors of three and one electrons, respectively, assuming in the case of copper a stable d^{10} configuration corresponding to Cu^+. The lithium atoms are one-electron donors by ionizing to Li^+. The stoichiometry $Al_{0.564}Cu_{0.116}Li_{0.32}$, coupled with the Audier model[41,43] for R-Al_5CuLi_3, leads to the stoichiometry $Al_{45}Cu_9^{26-}$ for the fundamental Samson complex structural unit corresponding to (45)(3) + (9)(1) + 26 = 170 electrons. In addition, application of the same Audier model to the cubic R phase of the Al–Cu–Li–Mg alloy of stoichiometry $Al_{0.52}Cu_{0.15}Li_{0.25}Mg_{0.08}$ leads to the stoichiometry $Al_{42}Cu_{12}Li_{20}Mg_6 = Al_{42}Cu_{12}^{32-}$, for its fundamental Samson complex structural unit also corresponding to (42)(3) + (12)(1) + 32 = 170 electrons. This suggests that 170 electrons is a "magic number" for the 54-atom Samson complex structural unit of cubic crystalline aluminum alloy phases closely related to the icosahedral quasicrystals of the i(AlZnMg) class. This model assumes that the aluminum and copper atoms in these structures occupy the vertices of the Samson complexes (layers a, c, and d in the Audier model), and the more electropositive metals (Li and Mg) occupy interstitial positions (layers b and e in the Audier model).

The topology of the 84-atom Samson complex (Figure 11–2) and the linkages of such complexes in a CsCl-type cubic lattice in R-Al_5CuLi_3 are consistent with the observed 170 electrons, corresponding to a closed shell electronic configuration with delocalized bonding in polyhedral cavities of the following two types:

1. The central $(Al,Cu)_{12}$ icosahedron (layer a) having the required (2)(12) + 2 = 26 skeletal electrons for globally delocalized chemical bonding topology similar to that in $B_{12}H_{12}^{2-}$ (see Section 4.3).
2. The twelve pentagonal pyramid $(Al,Cu)^a(Al,Cu)^b_{5/2}$ cavities (layers c and d), in which $(Al,Cu)^a$ refers to the single apical atom and $(Al,Cu)^b_{5/2}$ refers to the five basal atoms; each of the basal atoms is part of two different pentagonal pyramid cavities forming parts of different Samson complexes and thus has only two internal orbitals to contribute to the skeletal bonding of a given pentagonal pyramid cavity. In the view of the surface of the Samson complex in Figure 11–2, six of the pentagonal pyramid cavities can be seen.

The *nido* bonding in a pentagonal pyramid leading to $2n + 4 = 16$ skeletal electrons and eight bonding orbitals requires (6)(3) = 18 internal orbitals (see Section 4.4). However, in the pentagonal pyramid cavities in the Samson complexes in the R-Al_5CuLi_3 structure, there are only 3 + (5)(2) = 13 internal orbitals, since each of the five basal atoms can contribute only two internal orbitals (see above). For this reason, only one five-center core bond and four two-center surface bonds are possible using these 13 internal orbitals leading to 10 rather than 16 skeletal electrons for each pentagonal pyramid. The chemical bonding topology for a $(Al,Cu)_{12}(Al,Cu)_{12}(Al,Cu)_{60/2}$ Samson complex outlined in Table 11–2 can account for the observed 170 electrons while using the available 4 (12 + 12 + 60/2) = 216 valence orbitals of the sp^3 manifolds of the 54 vertex atoms.

TABLE 11–2
Electron and Orbital Counting in a Samson Complex Structural Unit in R-Al_5CuLi_3

	Total electrons	Total orbitals
(A) 1 central $(Al,Cu)_{12}$ icosahedron (layer a)		
Core bonding: (2 electrons; 12 orbitals) × 1 =	2 electrons	12 orbitals
Surface bonding: (12)(2) = (24 electrons; 24 orbitals)× 1 =	24 electrons	24 orbitals
(B) 12 two-center bonds between the external orbitals of the central $(Al,Cu)_{12}$ icosahedron (layer a) and the $(Al,Cu)_{12}$ icosahedron in t he next layer (layer b)		
(2 electrons; 2 orbitals) × 12	24 electrons	24 orbitals
(C) 12 $(Al,Cu)^a(Al,Cu)^b{}_{5/2}$ pentagonal pyramid cavities (layer d)		
Core bonding (5-center):		
(2 electrons; 5 orbitals) × 12 =	24 electrons	60 orbitals
Surface bonding : (4)(2) =		
(8 electrons; 8 orbitals) × 12 =	96 electrons	96 orbitals
Total electrons and orbitals per Samson complex:	*170 electrons*	*216 orbitals*

The above discussion considers a model for the structure of the crystalline cubic R-Al_5CuLi_3. Now consider possible perturbations in this model to give the quasicrystal T2-Al_5CuLi_3. In this connection, the 54-vertex Mackay icosahedron (Figure 11–6) appears as a structural unit in certain quasicrystals.[44] The Mackay icosahedron has a shell structure consisting of the following layers:

(a) a central icosahedron (not visible in Figure 11–6);
(b) a larger icosahedron formed from the external orbitals of the atoms in the central icosahedron (layer a) overlapping with an additional set of 12 atoms (black circles in Figure 11–6), which are located in the midpoints of the 12 pentagonal faces (6 visible) of the icosidodecahedron;
(c) a 30-vertex icosidodecahedron formed by placing atoms above each of the 30 edges of the larger icosahedron (layer b). Layer c is shown as white circles in Figure 11–6.

[44]P. Guyot, M. Audier, and R. Lequette, *J. Phys. (Paris)*, **47**, C3, 1986.

FIGURE 11–6. A view of the surface of a Mackay icosahedron showing the vertices of the larger icosahedron (layer b) as black circles and the vertices of the icosidodecahedron (layer c) as white circles. The vertices of the central icosahedron (layer a) are not visible.

Layers a and b of the Mackay icosahedron are identical to the first two layers of the Samson complex (i.e., layers a and c in the Audier model for R-Al_5CuLi_3), whereas the outer icosidodecahedron layer of the Mackay icosahedron (layer c) has exactly half the number of atoms of the outer truncated icosahedron in the Samson complex. Furthermore, the packing of the Samson complexes into the R-Al_5CuLi_3 lattice results in each of the peripheral truncated icosahedron atoms being shared between exactly two adjacent complexes (see above), so that a single Samson complex structural unit has the same 54 atoms as a corresponding Mackay icosahedron. This suggests a very close relationship between the packing of Samson complexes in the R-Al_5CuLi_3 crystal and a possible packing of Mackay icosahedra in a T2-Al_5CuLi_3 quasicrystal. In fact a concerted 90° rotation about a tangential axis of each of the 30 edges connecting pairs of pentagonal faces in the peripheral truncated icosahedron in each Samson complex of the R-Al_5CuLi_3 crystalline lattice converts a lattice of 54-atom Samson complexes into a lattice of 54-atom Mackay icosahedra.

11.4 ELEMENTAL CARBON CAGES: "FULLERENES"

Section 11.2 discusses some of the allotropes of elemental boron, namely α- and β-rhombohedral boron. Both of these boron allotropes have infinite solid state structures constructed from boron icosahedra. In the case of elemental carbon, two allotropes have been known since antiquity, namely graphite, which consists of infinite sheets of planar hexagons constructed from planar sp^2 hybridized carbon atoms, and diamond, an infinite three-dimensional structure constructed from tetrahedral sp^3 hybridized carbon atoms. One of the most exciting recent developments in chemistry has been the discovery of a third type of elemental carbon, namely the "fullerenes," which are finite cages of sp^2 carbon atoms containing only pentagonal and hexagonal faces.[45] Such fullerenes, named in honor of Buckminster Fuller because of their resemblance to geodesic domes, are *molecular* forms of elemental carbon, in contrast to the infinite diamond and graphite structures, which may be regarded as *polymeric* forms of elemental carbon. The most readily available fullerene is C_{60}, in which the 60 carbon atoms have been shown to be located at the vertices of a truncated icosahedron (see Figure 11–5) similar to the surface of a Samson complex (Figure 11–2). However, other larger, much less symmetrical fullerenes, such as C_{70}, are also known.

Schmalz, Klein, and their collaborators[46,47] have considered possible criteria for polyhedra forming stable carbon cages having no external groups and hence corresponding to allotropes of elemental carbon. The cages are considered to be constructed by bending planar carbon networks upon themselves in two directions. They thus consider polyhedra satisfying the following criteria:

1. A three-valent σ network corresponding to sp^2 carbon atoms with an extra p orbital to participate in delocalization leading to resonance stabilization;
2. A carbon cage homeomorphic to a sphere, i.e., no "doughnut holes", as in a torus;
3. All rings being pentagons and hexagons to minimize ring strain and non-aromatic rings;
4. Relatively high symmetry;
5. No pairs of pentagons sharing edges ("abutting five-sided rings") in order to avoid non-aromatic pentalene units;
6. Curvature spread uniformly over the cage in order to minimize strain.

These initial criteria are in accord with the following experimental observations in the mass spectra of carbon clusters:

1. An even number of vertices, i.e., C_{2n} is much more abundant than C_{2n-1} or C_{2n+1};
2. The smallest cage being the C_{60} truncated icosahedron, which is the most abundant in the mass spectra;
3. The next smallest cage being the C_{70} polyhedron, which is the next most abundant in the mass spectra.

Let us now consider some topological aspects of these criteria, using some of the ideas discussed in Section 1.4. The restriction to three σ bonds from

---

[45] H. W. Kroto, A. W. Allof, and S. P. Balin, *Chem. Rev.*, **91**, 1213, 1991.
[46] T. G. Schmalz, W. A. Seitz, D. J. Klein, and G. E. Hite, *J. Am. Chem. Soc.*, **110**, 1113, 1988.
[47] X. Liu, T. G. Schmalz, and D. J. Klein, *Chem. Phys. Lett.* **188**, 550, 1992.

each carbon vertex relates the number v of vertices to the number e of edges by the following equation which relates to Equation 1–13 :

$$2e = 3v \tag{11–12}$$

From Equation 1–10 and Euler's relationship (Equation 1–9), the following two equations can be derived in which f_n is the number of n-sided faces (or rings):

$$\sum_n n f_n = 2e \tag{11–13}$$

$$v + \sum_n f_n = e + 2 \tag{11–14}$$

Eliminating e and v from Equations 11–12, 11–13, and 11–14 gives a required balance between smaller and larger rings described by the following equation:

$$3f_3 + 2f_4 + f_5 = 12 + \sum_n (n - 6) f_n \tag{11–15}$$

Working towards criterion 3 above by setting $f_n = 0$ except for $n = 5$, 6, and 7 gives

$$f_5 = 12 + f_7 \tag{11–16}$$

and

$$f_6 = \tfrac{1}{2}v - 10 - 2f_7 \tag{11–17}$$

Setting f_6 and f_7 to zero in Equations 11–16 and 11–17 leads to the regular dodecahedron (Figure 4–1), which has the 12 pentagonal faces implied by Equation 11–16 and the 20 vertices implied by Equation 11–17. These conditions indicated in Equations 11–16 and 11–17 are necessary but not sufficient, in that there do not exist polyhedra for all possible sets of variables satisfying these equations. However, it is known[48] that for any nonnegative integer choice of f_7 there exists some values of f_6 for which a polyhedron is realizable and that if $f_7 = 0$, polyhedra are realizable for all f_6 other than 1. In order to give a more manageable list of polyhedra satisfying these conditions, the additional symmetry criterion (criterion 4 above) is introduced. If the symmetry of the cages is restricted to the "highest" possible symmetry,

[48]B. Grünbaum and T. S. Motzkin, *Can. J. Math.*, **15**, 744, 1963.

namely icosahedral, then each such polyhedron uniquely corresponds to a pair of integers h, k, such that $0 < h \geq k \geq 0$, with the number of vertices T being defined by the following equation:

$$T = 20(h^2 + hk + k^2) \qquad (11\text{–}18)$$

The simplest polyhedron satisfying these conditions again is the regular dodecahedron (Figure 4–1), for which $h = 1$ and $k = 0$, so that $T = 20$. The 60-vertex truncated icosahedron found in C_{60} (Figure 11–5) satisfies Equation 11–18 for $h = k = 1$. Polyhedra of icosahedral symmetry having only pentagonal and hexagonal faces in which all vertices have degree 3 and thus satisfying Equation 11–18 are called *Goldberg polyhedra*.[49] Fowler[50] has shown that the Goldberg polyhedra for carbon clusters follow an electron-counting rule analogous to the famous Hückel $4k + 2$ (k integer) rule (Section 4.3). Thus, when $h - k$ is divisible by three, the carbon cluster has a multiple of 60 atoms and has a closed shell electronic configuration; otherwise, the carbon cluster has $60m + 20$ atoms (m integer) and has an open shell electronic configuration. If a particular value of T satisfies Equation 11–18, then so do $3T$, $9T$, $27T$, etc. This follows from Equation 11–18 since if Equation 11–18 is satisfied, then

$$3T = H^2 + HK + K^2 \qquad (11\text{-}19)$$

where $H = h + 2k$ and $K = h - k$.

The relationship between h, k, H, K, T, and $3T$ is useful, since it has a geometric counterpart that provides a useful method for generating Goldberg polyhedra suitable for stable carbon cages. From Euler's relationship (Equation 1–9) and Equation 11–12, the following equation can be derived:

$$f = 2 + \frac{1}{2}v \qquad (11\text{–}20)$$

For a cage containing only pentagons and hexagons f_7 as well as f_n ($n \neq 5,6$) is zero so that from equations 11–16 and 11–17 one gets

$$f = 12 + f_6 \qquad (11\text{–}21)$$

where

$$f_6 = \frac{1}{2}v - 10 \qquad (11\text{–}22)$$

---

[49]M. Goldberg, *Tohoku Math. J.*, **43**, 104, 1937.

[50]P. W. Fowler, *Chem. Phys. Lett.*, **131**, 444 , 986.

On the other hand, a deltahedron has

$$e = \frac{3}{2}f \tag{11–23}$$

$$\text{so that } f = 2(v - 2). \tag{11–24}$$

If the deltahedron is the dual (Section 1.4) of a Goldberg polyhedron, it has 12 vertices of degree 5 and $v_6 = {}^1/_2{,}2)f - 10$ vertices of degree 6.

Now, to a Goldberg polyhedron with v_1 vertices let us apply a so-called "leapfrog" sequence, consisting of omnicapping followed by dualization. In the omnicapping stage, each face is capped so that the omnicapped polyhedron, which necessarily is a deltahedron, can be regarded as a "compound" of the original polyhedron and its dual. The omnicapped Goldberg polyhedron has $v_1 + {}^1/_2 v_1 + 2 = {}^1/_2 v_1 + 2$ vertices from Equation 11–20 and therefore by Equation 11–24 has $3v_1$ faces. Take the dual of the omnicapped Goldberg polyhedron. The result is a polyhedron with three times as many vertices as in the original Goldberg polyhedron. The processes of omnicapping and dualization both preserve the symmetry of the original polyhedron, so that the final polyhedron obtained by the leapfrog transformation has the icosahedral symmetry of the original Goldberg polyhedron and therefore is a new Goldberg polyhedron, with three times the number of vertices of the original Goldberg polyhedron. The effects of the leapfrog transformation on a pentagonal face, a hexagonal face, and a vertex of a Goldberg polyhedron is shown in Figure 11–7.

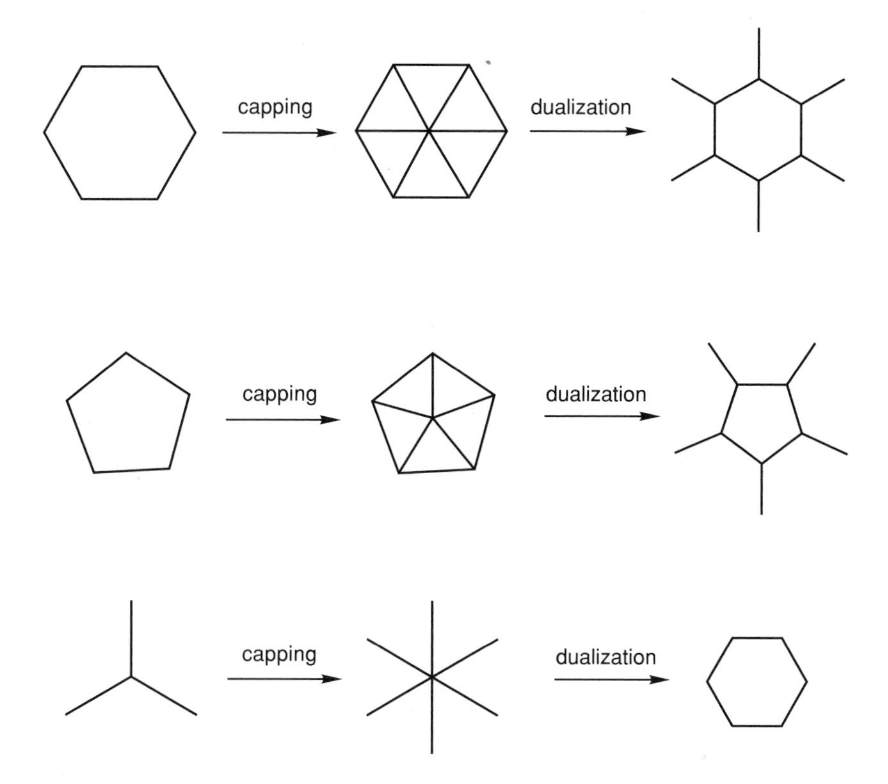

FIGURE 11–7. A schematic description of the effect of the leapfrog transformation on a hexagonal face, a pentagonal face, and a vertex of a Goldberg polyhedron with T vertices. The second column shows elements of the intermediate omnicapped deltahedron, and the third column their counterparts in the final Goldberg polyhedron with $3T$ vertices.

Chapter 12

POLYHEDRAL DYNAMICS

12.1 POLYHEDRAL ISOMERIZATIONS

The previous chapters of this book have been concerned with the static aspects of polyhedra relating to the configuration of atoms in a molecule at any specific time. This chapter is concerned with topological and graph-theoretical aspects of polyhedral dynamics relating to the changes in the configuration of atoms in a molecule as a function of time. The central concept in the study of polyhedral dynamics is that of a *polyhedral isomerization*. In this context, a polyhedral isomerization may be defined as a deformation of a specific polyhedron $\mathcal{P}_1$ until its vertices define a new polyhedron $\mathcal{P}_2$. Of particular interest are sequences of two polyhedral isomerization steps $\mathcal{P}_1 \rightarrow \mathcal{P}_2 \rightarrow \mathcal{P}_3$, in which the polyhedron $\mathcal{P}_3$ is combinatorially equivalent (Section 1.4) to the polyhedron $\mathcal{P}_1$ (i.e., the "same" polyhedron), although with some permutation of its vertices not necessarily the identity permutation. Such a polyhedral isomerization sequence $\mathcal{P}_1 \rightarrow \mathcal{P}_2 \rightarrow \mathcal{P}_3$ in which $\mathcal{P}_1$ and $\mathcal{P}_3$ are combinatorially equivalent may be called a *degenerate* polyhedral isomerization, with $\mathcal{P}_2$ as the *intermediate polyhedron*. Structures undergoing such degenerate isomerization processes are often called *fluxional*.[1] A degenerate polyhedral isomerization with a planar intermediate "polyhedron" (actually a polygon) may be called a *planar polyhedral isomerization*. The simplest example of a planar polyhedral isomerization is the interconversion of two tetrahedra ($\mathcal{P}_1$ and $\mathcal{P}_3$) through a square planar intermediate $\mathcal{P}_2$. Except for this simplest example, planar polyhedral isomerizations are unfavorable, owing to excessive intervertex repulsion.

Polyhedral isomerizations may be treated from either the macroscopic or microscopic points of view. The earliest work in this area, pioneered by the research of Muetterties[2,3,4,5] and Gielen,[6,7,8,9,10,11,12,13] as well as

---

[1]F. A. Cotton, *Acc. Chem. Res.*, **1**, 257, 1968.

[2]E. L. Muetterties, *J. Am. Chem. Soc.*, **90**, 5097, 1968.

[3]E. L. Muetterties, *J. Am. Chem. Soc.*, **91**, 1636, 1969.

[4]E. L. Muetterties, *J. Am. Chem. Soc.*, **91**, 4115, 1969.

[5]E. L. Muetterties and A. T. Storr, *J. Am. Chem. Soc.*, **91**, 3098, 1969.

[6]M. Gielen and J. Nasielski, *Bull. Soc. Chim. Belges*, **78**, 339, 1969.

[7]M. Gielen and J. Nasielski, *Bull. Soc. Chim. Belges*, **78**, 351, 1969.

[8]M. Gielen, C. Depasse-Delit, and J. Nasielski, *Bull. Soc. Chim. Belges*, **78**, 357, 1969

[9]M. Gielen and C. Depasse-Delit, *Theor. Chim. Acta*, **14**, 212, 1969.

[10]M. Gielen, G. Mayence, and J. Topart, *J. Organometal. Chem.*, **18**, 1, 1969.

[11]M. Gielen, M. de Clercq, and J. Nasielski, *J. Organometal. Chem.*, **18**, 217, 1969.

[12]M. Gielen and N. Vanlautem, *Bull. Soc. Chim. Belges*, **79**, 679, 1970.

[13]M. Gielen, *Bull. Soc. Chim. Belges*, **80**, 9, 1971.

Musher,[14,15] Klemperer,[16,17,18] and Brocas[19], focussed on the *macroscopic* picture, namely relationships between different permutational isomers. Such relationships may be depicted by graphs called *topological representations*, in which the vertices correspond to different permutational isomers and the edges correspond to single degenerate polyhedral isomerization steps. Subsequent work treats the *microscopic* picture, in which the details of polyhedral topology are used to elucidate possible single polyhedral isomerization steps, namely which types of isomerization steps are possible. Both approaches will be treated in this chapter.

Consider an ML_n coordination compound having n ligands, or a cluster compound having n vertices. There are a total of $n!$ permutations of the sites occupied by these ligands or vertices which form the symmetric group (Section 2.5). The symmetric group P_n is the automorphism ("symmetry") group of the corresponding complete graph K_n (Section 1.2).

Now consider the symmetry point group G (or more precisely, the framework group—Section 2.6) of the above ML_n coordination compound or n-vertex cluster compound. This group has $|G|$ operations, of which $|R|$ are proper rotations: $|G|/|R| = 2$ if the compound is achiral and $|G|/|R| = 1$ if the compound is chiral. The $n!$ distinct permutations of the n sites in the coordination compound or cluster are divided into $n!/|\mathrm{R}|$ right cosets, which represent the permutational isomers since the permutations corresponding to the $|R|$ proper rotations of a given isomer do not change the isomer but merely rotate it in space. This leads naturally to the concept of *isomer count*, I, namely

$$I = n!/|R| \qquad (12\text{–}1)$$

if all vertices are distinguishable. Similarly the quotient

$$E = n!/|G| = I/2 \qquad (12\text{–}2)$$

for a given *chiral* polyhedron corresponds to the number of enantiomeric pairs. The isomer count I indicates the complexity of macroscopic models for polyhedral isomerizations, such as the topological representations discussed in Section 12.3.

---

[14]J. I. Musher, *J. Am. Chem. Soc.*, **94**, 5662, 1972.
[15]J. I. Musher, *Inorg. Chem.*, **11**, 2335, 1972.
[16]W. G. Klemperer, *J. Chem. Phys.*, **56**, 5478, 1972.
[17]W. G. Klemperer, *J. Am. Chem. Soc.*, **94**, 6940, 1972.
[18]W. G. Klemperer, *J. Am. Chem. Soc.*, **94**, 8360, 1972.
[19]J. Brocas, *Top. Curr. Chem.*, **32**, 43, 1972.

12.2 MICROSCOPIC MODELS: DIAMOND-SQUARE-DIAMOND PROCESSES IN DELTAHEDRA

Microscopic approaches to polyhedral rearrangements dissect such processes into elementary steps. The most important such elementary step in deltahedral rearrangements is the *diamond-square-diamond* process (Figure 12–1), which was first recognized in a chemical context by Lipscomb in 1966[20] as a generalization of a process earlier proposed by Berry[21] for rearrangements of five-vertex trigonal bipyramids. Such a diamond-square-diamond process (or "dsd process") in a polyhedron occurs at two triangular faces sharing an edge (Figure 12–1). In this process, a configuration such as p_1 can be called a *dsd situation*, and the edge AB can be called a *switching edge*. If a, b, c, and d are taken to represent the degrees of the vertices A, B, C, and D, respectively, in p_1, then the *dsd type* of the switching edge AB can be represented as $ab(cd)$. In this designation, the first two digits refer to the degrees of the vertices joined by AB but contained in the faces (triangles) having AB as the common edge (i.e., C and D in p_1). The quadrilateral face formed in structure p_2 may be called a *pivot face*.

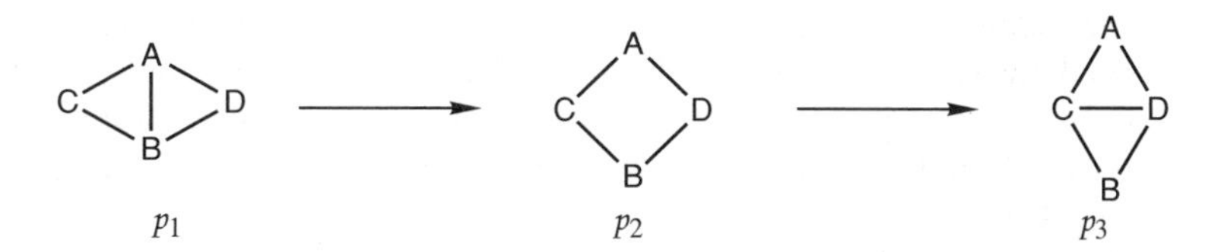

FIGURE 12–1. A schematic picture of the diamond-square-diamond process occurring at a pair of triangular faces joined by an edge.

Consider a deltahedron having e edges. Such a deltahedron has e distinct dsd situations, one corresponding to each of the e edges acting as the switching edge. Applications of the dsd process at each of the dsd situations in a given deltahedron leads in each case to a new deltahedron. In some cases, the new deltahedron is identical to the original deltahedron. In such cases, the switching edge can be said to be *degenerate*. A dsd process involving a degenerate switching edge represents a pathway for a degenerate polyhedral isomerization of the deltahedron. A deltahedron having one or more degenerate edges is inherently fluxional, whereas a deltahedron without degenerate edges is inherently rigid.

The dsd type of a degenerate edge $ab(cd)$ can be seen by application of the process $p_1 \rightarrow p_2 \rightarrow p_3$ to satisfy the following conditions:

$$c = a - 1 \text{ and } d = b - 1 \text{ or } c = b - 1 \text{ and } d = a - 1 \qquad (12\text{–}3)$$

[20]W. N. Lipscomb, *Science*, **153**, 373, 1966.

[21]R. S. Berry, *J. Chem. Phys.*, **32**, 933, 1960.

Using these conditions, the chemically significant deltahedra depicted in Figure 4–5 can be very easily checked for the presence of one or more degenerate edges with the following results:

1. **Tetrahedron.** No dsd process of any kind is possible since the tetrahedron is the complete graph K_4. A tetrahedron is therefore inherently rigid.
2. **Trigonal bipyramid.** The three edges connecting pairs of equatorial vertices are degenerate edges of the type 44(33). A dsd process using one of these degenerate edges as the switching edge and involving a square pyramid intermediate corresponds to the Berry pseudorotation,[21,22]which is believed to be the mechanism responsible for the stereochemical nonrigidity of trigonal bipyramidal complexes, even at relatively low temperatures.[23]
3. **Octahedron.** The highly symmetrical octahedron has no degenerate edges and is therefore inherently rigid.
4. **Pentagonal bipyramid.** The pentagonal bipyramid has no degenerate edges and, thus, by definition is inherently rigid. However, a dsd process using a 45(44) edge of the pentagonal bipyramid (namely an edge connecting an equatorial vertex with an axial vertex) gives a capped octahedron (Figure 6–2). The capped octahedron is a low energy polyhedron for ML_7 coordination complexes[24] but a forbidden polyhedron for boranes and carboranes because of its tetrahedral chamber.[25]
5. **Bisdisphenoid.** The eight-vertex bisdisphenoid has four pairwise degenerate edges, which are those of the type 55(44) located in the subtetrahedron consisting of the degree 5 vertices of the bisdisphenoid (see Figure 4–5, which has the vertex degrees labeled). Thus two successive or more likely concerted (parallel) dsd processes involving opposite 55(44) edges (i.e., a pair related by a C_2 symmetry operation) convert one bisdisphenoid into another bisdisphenoid through a square antiprismatic intermediate. Thus, a bisdisphenoid, like the trigonal bipyramid discussed above, is inherently fluxional.
6. **4,4,4-Tricapped Trigonal Prism.** The three edges of the type 55(44), corresponding to the "vertical" edges of the trigonal prism, are degenerate. A dsd process using one of these degenerate edges as the switching edge involves a C_{4v} 4-capped square antiprism intermediate. Nine-vertex systems are therefore inherently fluxional.
7. **4,4-Bicapped Square Antiprism.** This polyhedron has no degenerate edges and therefore is inherently rigid.

---

[22]R. R. Holmes, *Acc. Chem. Res.*, **5**, 296, 1972.
[23]P. C. Lauterbur and F. Ramirez, *J. Am. Chem. Soc.*, **90**, 6722, 1968.
[24]D. L. Kepert, *Prog. Inorg. Chem.*, **25**, 41, 1979.
[25]R. B. King and D. H. Rouvray, *J. Am. Chem. Soc.*, **99**, 7834, 1977.

8. **Edge-coalesced Icosahedron.** The four edges of the type 56(45) are degenerate. This eleven-vertex deltahedron is therefore inherently fluxional.
9. **Icosahedron.** This highly symmetrical polyhedron, like the octahedron, has no degenerate edges and is therefore inherently rigid.

This simple analysis indicates that in deltahedral structures the 4, 6, 10, and 12 vertex structures are inherently rigid; the 5, 8, 9, and 11 vertex structures are inherently fluxional; and the rigidity of the 7 vertex structure depends upon the energy difference between the two most symmetrical seven-vertex deltahedra, namely the pentagonal bipyramid and the capped octahedron. This can be compared with experimental fluxionality observations by boron-11 nuclear magnetic resonance on the deltahedral borane anions $B_nH_n^{2-}$ ($6 \leq n \leq 12$),[26] where the 6, 7, 9, 10, and 12 vertex structures are found to be rigid, and the 8 and 11 vertex structures are found to be fluxional. The only discrepancy between experiments and these very simple topological criteria for fluxionality arises in the nine vertex structure $B_9H_9^{2-}$.

These topological criteria for fluxionality are not sufficient for a structure to be fluxional, since degenerate isomerizations of inherently fluxional structures may be forbidden because of orbital symmetry. For example, Gimarc and Ott have done orbital symmetry analyses on dsd processes in the carboranes $C_2B_{n-2}H_n$. Thus, in both the five-vertex trigonal bipyramidal carboranes[27] and the nine-vertex tricapped trigonal prismatic boranes and carboranes,[28] the degenerate single dsd processes suggested by the inherent fluxionality of these deltahedra are blocked by crossings of filled and vacant molecular orbitals. However, a double dsd process in the tricapped trigonal prism is orbital symmetry allowed. The symmetry-forbidden nature of the single dsd process in the tricapped trigonal prism suggests an explanation of the experimentally observed rigidity of tricapped trigonal prismatic $B_9H_9^{2-}$, in contrast to the inherent fluxionality of the polyhedron. The Gimarc-Ott analysis cannot be extended directly to transition metal clusters involving *d* orbitals of the vertex atoms and is not applicable to coordination rather than cluster polyhedra.

Mingos and Johnston[29] have related the orbitally allowed and forbidden nature of degenerate dsd processes in deltahedra to the symmetry of the intermediate polyhedron. Thus, the symmetry-forbidden natures of the degenerate single dsd processes for the five-vertex trigonal bipyramid and the nine-vertex tricapped trigonal prism are consequences of the C_{4v} symmetry of the square pyramid and capped square antiprism intermediates (Figure 8–1), respectively, corresponding to a pseudo 90° rotation of the deltahedron during

[26] R. B. King, *Inorg. Chim. Acta*, **49**, 237, 1981.

[27] B. M. Gimarc and J. J. Ott, *Inorg. Chem.*, **25**, 83, 1986.

[28] B. M. Gimarc and J. J. Ott, *Inorg. Chem.*, **25**, 2708, 1986.

[29] D. M. P. Mingos and R. L. Johnston, *Polyhedron*, **7**, 2437, 1988.

its degenerate isomerization. However, if a degenerate dsd process proceeds through a C_{2v} or C_s intermediate, it corresponds to a pseudoreflection of the deltahedron, and the process is symmetry allowed. Thus, the single dsd processes for the inherently fluxional eight-vertex bisdisphenoid and eleven-vertex edge-coalesced icosahedron are symmetry-allowed, in accord with the boron-11 nuclear magnetic resonance data on $B_8H_8^{2-}$ and $B_{11}H_{11}^{2-}$ mentioned above.

12.3 MACROSCOPIC MODELS: TOPOLOGICAL REPRESENTATIONS OF POLYHEDRAL ISOMERIZATIONS

Topological representations are graphs describing the relationships between the different permutational isomers of a given polyhedron. In such a graph, the vertices correspond to isomers and the edges correspond to isomerization steps. The number of vertices corresponds to the isomer count, $I = n!/|R|$ (Equation 12–1). The degree of a vertex corresponds to the number of new permutational isomers generated from the isomer represented by the vertex in a single step; this is called the *connectivity*, δ, of the vertex.

A simple example of a topological representation is provided in the four-vertex system by the degenerate planar isomerization of a tetrahedron into its enantiomer through a square planar intermediate. The isomer count for the tetrahedron, I_{tet}, is $4!/|T| = 24/12 = 2$, and the isomer count for the square, I_{sq}, is $4!/|D_4| = 24/8 = 3$. A topological representation of this process is a $K_{2,3}$ bipartite graph (Section 1.2), which is derived from the trigonal bipyramid by deletion of the three equatorial-equatorial edges (Figure 12–2). The two axial vertices correspond to the two tetrahedral isomers, and the three equatorial vertices correspond to the three square planar isomers. The connectivities of the tetrahedral (δ_{tet}) and square planar (δ_{sq}) isomers are 3 and 2, respectively, in accord with the degrees of the corresponding vertices of the $K_{2,3}$ graph (Figure 12–2). Thus $I_{tet}\delta_{tet} = I_{sq}\delta_{sq} = 6$; this is an example of the *closure* condition $I_a\delta_a = I_b\delta_b$ required for a topological representation with vertices representing more than one type of polyhedron.

Some interesting graphs are found in the topological representations of the five-vertex trigonal bipyramidal structures. The trigonal bipyramid has an isomer count $I = 5!/|D_3| = 120/6 = 20$, corresponding to 10 enantiomeric pairs. A given trigonal bipyramidal isomer can be described by the labels of its two axial positions (i.e., the single pair of vertices not connected by an edge) with a bar used to distinguish enantiomers. In a single degenerate dsd isomerization of a trigonal bipyramid through a square pyramid intermediate, both axial vertices of the original trigonal bipyramid become equatorial vertices in the new trigonal bipyramid, leading to a connectivity of three for dsd isomerizations of trigonal bipyramids. The corresponding topological representation thus is a 20 vertex graph in which each vertex has degree 3. However, additional properties of dsd isomerizations of trigonal bipyramids

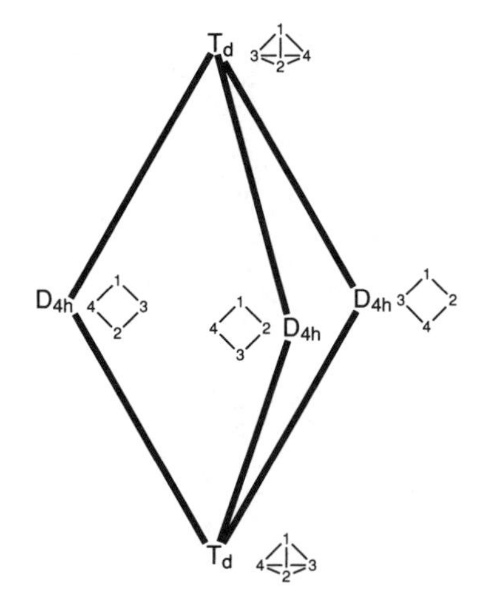

FIGURE 12–2. The $K_{2,3}$ bipartite graph as a topological representation of the degenerate planar isomerization of a tetrahedron (T_d) into its enantiomer through a square planar intermediate (D_{4h}). The isomers corresponding to the vertices of the $K_{2,3}$ bipartite graph are depicted next to the vertex labels.

exclude the regular (I_h) dodecahedron as a topological representation unless double group form is used to produce pseudohexagonal faces. A graph suitable for the topological representation of dsd isomerizations of trigonal bipyramids is the Desargues-Levy graph, depicted in Figure 12–3.

Less complicated but still useful topological representations can be obtained by using each vertex of the graph to represent a set of isomers, provided that each vertex represents sets of the same size and interrelationship and each isomer is included in exactly one set. A simple example is the use of the Petersen's graph (Figure 12–3 bottom) as a topological representation of isomerizations of the 10 trigonal bipyramid enantiomer pairs ($E = 5!/|D_{3h}| = 120/12 = 10$) by dsd processes. The use of Petersen's graph for this purpose relates to its being the odd graph O_3; an *odd graph* O_k is defined as follows[30]: its vertices correspond to subsets of cardinality $k - 1$ of a set S of cardinality $2k - 1$, and two vertices are adjacent if and only if the corresponding subsets are disjoint.

Useful topological representations can also be obtained for six-vertex systems. Here the process of interest is the degenerate triple dsd isomerization of the octahedron through a trigonal prismatic intermediate, which is the underlying topology of both the Bailar[31] and Ray and Dutt[32] twists for octahedral M(bidentate)$_3$ chelates. The isomer counts are $I_{oct} = 6!/|O| = 720/24 = 30$ for the octahedron and $I_{tp} = 6!|D_3| = 720/6 = 120$ for the trigonal prism. A pentagonal (I_h) dodecahedron in double group form can serve for the

[30]N. L. Biggs, *Algebraic Graph Theory*, Cambridge University Press, London, 1974.

[31]J. C. Bailar, Jr., *J. Inorg. Nucl. Chem.*, **8**, 165, 1985.

[32]P. Ray and N. K. Dutt, *J. Indian Chem. Soc.*, **20**, 81, 1943.

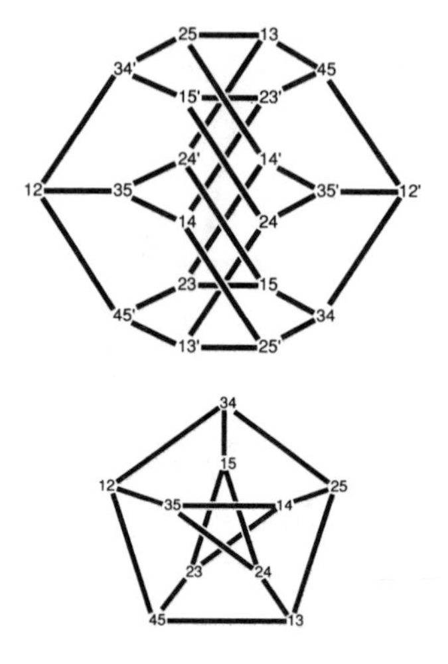

FIGURE 12–3. Topological representation of the isomerizations of trigonal bipyramids through dsd processes (Berry pseudorotations). The two digits represent labels for the axial positions, with primes used to indicate enantiomers. Top: Desargues-Levy graph as a topological representation for dsd isomerizations of the 20 trigonal bipyramid isomers. Bottom: Petersen's graph as a topological representation for dsd isomerizations of the 10 enantiomer pairs of the trigonal bipyramid isomers.

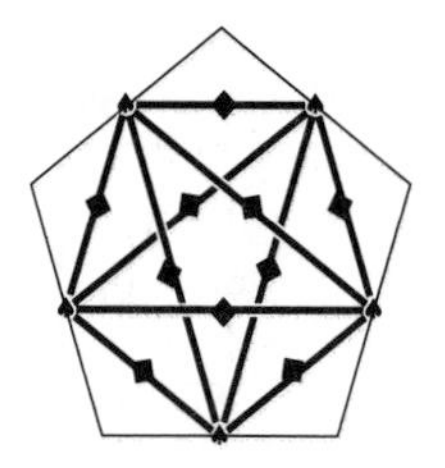

FIGURE 12–4. One of the twelve pentagonal faces of the I_h double group (pentagonal dodecahedron), used as a topological representation for the degenerate triple dsd isomerization of the octahedron (O_h) through a trigonal prismatic intermediate (D_{3h}). The five spades (♠) at the midpoints of the sides of the face represent five of the 30 octahedron permutational isomers, whereas the 10 diamonds (♦) at the midpoints of the edges of the K_5 graph drawn on the face represent 10 of the 120 trigonal prism permutational isomers.

topological representation for this process. The midpoints of the 30 edges of the dodecahedron are the 30 octahedron isomers. Line segments across a pentagonal face connecting these edge midpoints correspond to triple dsd isomerization processes; the midpoints of these lines correspond to the 120 trigonal prismatic isomers, with 10 such isomers being located in each of the 12 faces of the pentagonal dodecahedron. Figure 12–4 depicts one of the 12 pentagonal faces of this double group pentagonal dodecahedral representation; the ten lines on this face representing isomerization processes form a K_5 graph. This system is closed, since the connectivities of the octahedron (δ_{oct}) and trigonal prism (d_{tp}) are 8 and 2, respectively, leading to the closure relationship $I_{oct}\delta_{oct} = I_{tp}\delta_{tp} = 240$.

Development of topological representations for systems having more than six vertices is complicated by intractably large isomer counts. Thus, the isomer count of the seven-vertex polyhedron with the largest number of symmetry elements, namely the pentagonal bipyramid, is $7!/|D_5| = 5040/10 = 504$. Similarly, the isomer counts of the cube, hexagonal bipyramid, square antiprism, and bisdisphenoid are 40320/24 = 1680, 40320/12 = 3360, 40320/8 = 5040, and 40320/4 = 10080, respectively. Graphs corresponding to topological representations involving such large numbers of polyhedral isomers are clearly unwieldy and unmanageable, even though a recent paper by Klin, Tratch, and Zefirov[33] suggests some group-theoretical methods for studying topological representation graphs having large numbers of vertices.

The problem of representing permutational isomerizations in seven- and eight-vertex polyhedra can be simplified if subgroups of the symmetric groups P_n (n = 7, 8) can be found which contain all of the symmetries of all of the polyhedra of interest. However, this is *not* possible for the seven-vertex system, since there is no subgroup of P_7 that contains both the five-fold symmetry of the pentagonal bipyramid and the three-fold symmetry of the capped octahedron. However, the situation with the eight-vertex system is more favorable, since the wreath product group[34,35,36] $P_4[P_2]$ of order 384 contains all of the symmetries of the cube, hexagonal bipyramid, square antiprism, and bisdisphenoid,[37] which are all of the eight-vertex polyhedra of actual or potential chemical interest. The major effect of reducing the symmetry by a factor of 105 ($3 \times 5 \times 7$) in going from P_8 to $P_4[P_2]$ is the deletion of five-fold and seven-fold symmetry elements. Such symmetry elements are not of interest in this context since none of the 257 eight-vertex polyhedra has five-fold symmetry elements[38,39] and the only eight-vertex polyhedron having a seven-fold symmetry element is the heptagonal pyramid, which is not of interest in this particular chemical context. Restricted isomer counts $I^* = 384/|R|$, based on subgroups of the wreath product group $P_4[P_2]$ rather than the symmetric group P_8, are the more manageable numbers 16, 32, 48, and 96 for the cube, hexagonal bipyramid, square antiprism, and bisdisphenoid, respectively.

The concept of restricting vertex permutations in eight-vertex systems to the wreath product group $P_4[P_2]$ rather than the fully symmetric P_8 group can be restated in graph-theoretical terms using the hyperoctahedral graph H_4.[30] Therefore, such a restriction of permutations from P_8 to $P_4[P_2]$ can be called a

---

[33] M. H. Klin, S. S. Tratch, and N. S. Zefirov, *J. Math. Chem.*, **7**, 135 ,1 991.

[34] G. Pólya, *Acta Math.*, **68**, 143, 1937.

[35] N. Debruijn in *Applied Combinatorial Mathematics*, E. F. Beckenbach, ed., John Wiley & Sons, New York, 1964.

[36] J. G. Nourse and K. Mislow, *J. Am. Chem. Soc.*, **97**, 4571, 1975.

[37] R. B. King, *Inorg. Chem.*, **20**, 363, 1981.

[38] P. J. Federico, *Geom. Ded.*, **3**, 469 , 975.

[39] D. Britton and J. Dunitz, *Acta Cryst.*, **A29**, 362, 1973.

hyperoctahedral restriction. The hyperoctahedral graphs underlying this restriction are designated as H_n and have $2n$ vertices and $2n(n - 1)$ edges, with every vertex connected to all except one of the remaining vertices, so that each vertex of H_n has degree $2(n - 1)$. The name "hyperoctahedral" comes from the fact that an H_n graph is the 1-skeleton of the analogue of the octahedron (called the "cross-polytope") in n-dimensional space.[40] The hyperoctahedral graphs H_2 and H_3 thus correspond to the square and octahedron, respectively. The $P_4[P_2]$ wreath product group is the automorphism ("symmetry") group of the hyperoctahedral graph H_4, just as the P_8 symmetric group is the automorphism group of the complete graph K_8. The hyperoctahedral graph H_4 is the 1-skeleton of the four-dimensional cross-polytope, which is the dual of the tesseract, the four-dimensional analogue of the cube; a two-dimensional projection of the tesseract is depicted in Figure 12–5. The tesseract has eight three-dimensional cubic "cells." In the two-dimensional projection of four-dimensional space in Figure 12–5, these eight cubic "cells" correspond to the "inner" cube, six cubes each containing one of the six faces of the "outer" cube and one of the six faces of the "inner" cube, and a cube containing the six faces of the "outer" cube; this last cubic "cell" consists of the three-dimensional space "outside" the "outer" cube and includes the point at infinity.

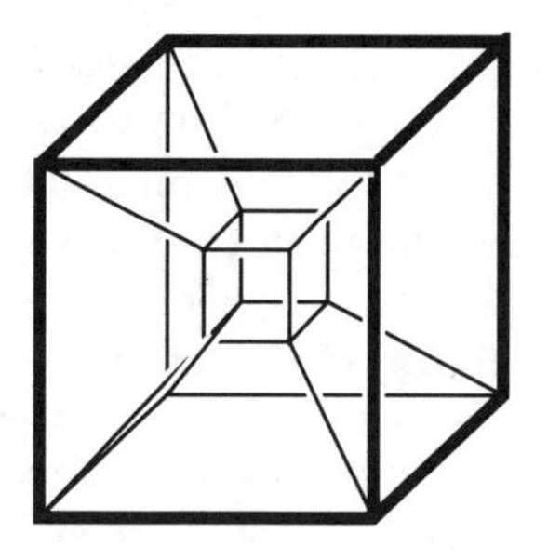

FIGURE 12–5. The tesseract or "four-dimensional cube."

Using these ideas, topological representations for isomerizations of eight-vertex polyhedra are depicted in Figures 12–6 and 12–7. Vertex and edge midpoints in these representations correspond to the $E^* = 384/|G|$ hyperoctahedrally restricted enantiomer pairs (E^* = 8, 16, 24, and 48 for the cube, hexagonal bipyramid, square antiprism, and bisdisphenoid, respectively), except, because of the hyperoctahedral reduction in symmetry, the number of points for the square antiprism must be doubled.[41] Figure 12–6 is a $K_{4,4}$ bipartite graph (Section 1.2), in which the 8 cube enantiomer pairs are located in the centers of the hexagons and the 16 hexagonal bipyramid enantiomer pairs are located at the edge midpoints. Since both the

---

[40]B. Grünbaum, *Convex Polytopes*, Interscience, New York, 1967.

[41]R. B. King, *Theor. Chim. Acta*, **59**, 25, 1981.

cube and hexagonal bipyramid are forbidden polyhedra (i.e., cannot be formed using only *s*, *p* and *d* orbitals),[42] this portion of the topological representation for hyperoctahedrally restricted eight-vertex systems is *not* accessible if only *s*, *p* and *d* orbitals are available for chemical bonding.

The detailed structure of a hexagon corresponding to a given pair of cube enantiomers is depicted in Figure 12–7. The vertices of the hexagon correspond to the square antiprisms that can be generated from the cube in the center by twisting opposite pairs of faces. The midpoints of the hexagon edges correspond to bisdisphenoid enantiomer pairs. Traversing the circumference of a given hexagon corresponds to a sequence of double dsd processes, interconverting the bisdisphenoids located at the midpoints of the two joined hexagonal edges meeting at a vertex through the square antiprism intermediate represented by the vertex joining the edges. Since both the bisdisphenoid and square antiprism can be formed using only *s*, *p*, and *d* orbitals (see Section 2.4), the circumference of the hexagon is accessible in ML_8 systems in which the central atom M has the usual sp^3d^5 nine-orbital manifold. Thus, in the usual situation *not* involving *f* orbitals, isomerizations are restricted to the circumference of a given hexagon in Figure 12–6 and cannot occur by moving from one hexagon to another.

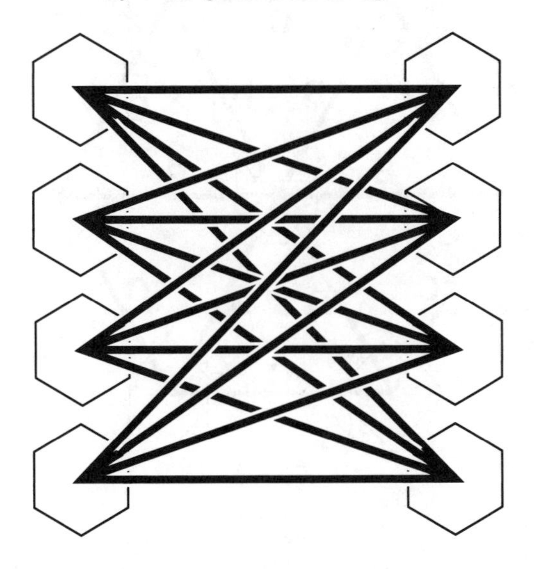

FIGURE 12–6. The $K_{4,4}$ bipartite graph topological representation of the hyperoctahedrally restricted permutational isomerizations involving the cube, hexagonal bipyramid, square antiprism and the bisdisphenoid. The hexagons represent the eight cube hyperoctahedrally restricted enantiomeric pairs; the 16 edges connecting the hexagons (i.e., the midpoints of the edges of the $K_{4,4}$ bipartite graph) represent the 16 hexagonal bipyramid hyperoctahedrally restricted enantiomeric pairs; the 48 vertices of the eight hexagons represent 48 isomeric square antiprisms; and the midpoints of the 48 edges of the eight hexagons represent the 48 bisdisphenoid hyperoctahedrally restricted enantiomeric pairs.

---

[42] R. B. King, *Theor. Chim. Acta*, **64**, 453, 1984.

12.4 GALE TRANSFORMATIONS AND GALE DIAGRAMS

Gale diagrams provide an elegant method for the reduction of the degrees of freedom of allowed vertex motions in polyhedral rearrangements of certain types. The underlying mathematics was developed by mathematicians for studying the properties of polytopes (the general term for polyhedra in any number of dimensions) in d-dimensional space having only a few vertices more than the minimum $d + 1$ vertices. In a chemical context, they can be used to study possible rearrangements of six-atom structures by depicting skeletal rearrangements of six atoms as movements of six points on the circumference of a circle or from the circumference to the center of the circle, subject to severe restrictions that reduce such possible movements to a manageable number.[43]

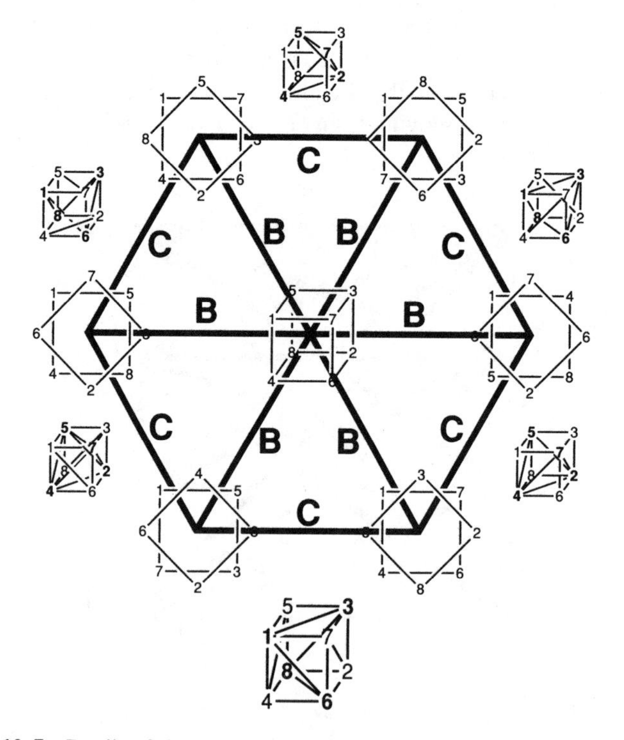

FIGURE 12–7:. Details of the hexagon in the topological representation in Figure 12–6. The spokes B represent quadruple dsd square antiprism-cube-square antiprism isomerizations, and the edges C represent double dsd bisdisphenoid-square antiprism-bisdisphenoid isomerizations, with a cube in the center of the hexagon, six isomeric square antiprisms at the vertices of the hexagon, and six isomeric bisdisphenoids at the midpoints of the edges of the hexagon. The bisdisphenoids are depicted as the topologically equivalent cubes, with six diagonals with the numbers of the degree 5 vertices indicated in bold face.

---

[43] R. B. King, *J. Mol. Struct. THEOCHEM*, **185**, 15, 1989.

Consider a polytope $\mathcal{P}$ in d-dimensional space $\Re^d$. The minimum number of vertices of such a polytope is $d + 1$, and there is only one such polytope,, namely the d-simplex.[30] The combinatorially distinct possibilities for polytopes having only $d + 2$ and $d + 3$ vertices (polytopes with "few" vertices) are also rather limited, and through a Gale transformation[44] can be represented faithfully in a space of less than d dimensions. More specifically, if $\mathcal{P}$ is a d-dimensional polytope with v vertices, a Gale transformation leads to a Gale diagram of $\mathcal{P}$, consisting of v points in $(v-d-1)$-dimensional space $\Re^{d-1}$ in *one-to-one correspondence* with the vertices of $\mathcal{P}$., From the Gale diagram it is possible to determine all of the combinatorial properties of $\mathcal{P}$ such as the subsets of the vertices of $\mathcal{P}$ that define faces of $\mathcal{P}$, the combinatorial types of these faces, etc. Of particular significance in the present context is the fact that the combinatorial properties of a polytope $\mathcal{P}$ which can be determined by the Gale diagram include all possible isomerizations (rearrangements) of $\mathcal{P}$ to other polytopes having the same number of vertices and imbedded in the same number of dimensions as $\mathcal{P}$. Also of particular importance is the fact that, if v is not much larger than d (i.e., if $v \leq 2d$), then the dimension of the Gale diagram is smaller than that of the original polytope $\mathcal{P}$.

Now, consider polyhedra in the ordinary three-dimensional space of interest in chemical structures (i.e., $d = 3$). Gale diagrams of five- and six-vertex polyhedra can be imbedded into one- or two-dimensional space, respectively, thereby simplifying analysis of their possible vertex motions leading to non-planar polyhedral isomerizations of these polyhedra of possible interest in a chemical context.

In order to obtain a Gale diagram for a given polyhedron, the polyhedron is first subjected to a *Gale transformation*. Consider a polyhedron with v vertices as a set of v points $X_1,\ldots, X_v$ in three-dimensional space $\Re^3$. These points may be regarded as three-dimensional vectors, $\mathbf{X}_n = (x_{n,1}, x_{n,2}, x_{n,3})$, $1 \leq n \leq v$, from the origin to the vertices of the polyhedron. In addition, consider a set of points $\mathcal{D}(\mathrm{A})$ in v-dimensional space $\Re^v$, $\mathbf{A} = (a_1,\ldots,a_v)$, such that the following sums vanish:

$$\sum_{i=1}^{v} a_i x_{i,k} = 0 \text{ for } 1 \leq k \leq 3 \qquad (12\text{–}4a)$$

$$\sum_{i=1}^{v} a_i = 0 \qquad (12\text{-}4b)$$

[44]P. McMullen and G. C. Shephard, *Convex Polytopes and the Upper Bound Conjecture*, Cambridge University Press, Cambridge, England, 1971.

Equation 12–4a may also be viewed as three orthogonality relationships between the v-dimensional vector $\mathbf{A} = (a_1,\ldots,a_v)$ and the three v-dimensional vectors $(x_{1,k}, x_{2,k},\ldots, x_{v,k})$, $1 \le k \le 3$. Now, consider the locations of the vertices of the polyhedron as the following $v \times 4$ matrix:

$$\mathbf{D}_0 = \begin{pmatrix} x_{1,1} & x_{1,2} & x_{1,3} & 1 \\ x_{2,1} & x_{2,2} & x_{2,3} & 1 \\ \cdot & \cdot & \cdot & \\ \cdot & \cdot & \cdot & \\ x_{v,1} & x_{v,2} & x_{v,3} & 1 \end{pmatrix} \tag{12–5}$$

Consider the columns of $\mathbf{D}_0$ as vectors in $\Re^v$. Since $\mathbf{D}_0$ has rank 4, the four columns of $\mathbf{D}_0$ are linearly independent. Hence, the subspace $\mathcal{M}(\mathrm{X})$ of $\Re^v$ represented by these four linearly independent columns has dimension 4. Its orthogonal complement $\mathcal{M}(\mathbf{A})^{\perp} = \{\mathbf{A} \in \Re^v \mid \mathbf{A}\cdot\mathbf{X} = 0 \text{ for all } \mathbf{X} \in \mathcal{M}(\mathbf{X})\}$ coincides with $\mathcal{D}(\mathbf{A})$, defined above by Equations 12–4a and 12–4b. Therefore,

$$\dim \mathcal{D}(\mathbf{A}) = \dim \mathcal{M}(\mathbf{A})^{\perp} = v - \dim \mathcal{M}(\mathbf{X}) = v - 4 \tag{12–6}$$

Now define the following $v \times (v–4)$ matrix:

$$\mathbf{D}_1 = \begin{pmatrix} a_{1,1} & a_{1,2} & \cdot & \cdot & \cdot & a_{1,v-4} \\ a_{2,1} & a_{2,2} & \cdot & \cdot & \cdot & a_{2,v-4} \\ \cdot & \cdot & & & & \cdot \\ \cdot & \cdot & & & & \cdot \\ a_{v,1} & a_{v,2} & \cdot & \cdot & \cdot & a_{v,v-4} \end{pmatrix} \tag{12–7}$$

The v rows of $\mathbf{D}_1$ may be considered as vectors in $(v–4)$-dimensional space; conventionally the jth row is denoted by $\bar{x}_j = (a_{i,1}, a_{j,2},\ldots,a_{j,v-4})$ for $j = 1,\ldots,v$.

The final result of this construction is the assignment of a point $\bar{x}_j$ in (v–4)- dimensional space ($\Re^{v-4}$) to each vertex x_j of the polyhedron. The collection of v points $\bar{x}_1,\ldots\bar{x}_v$ in $\Re_{v-4}$ is called a *Gale transform* of the set of vertices $x_1,\ldots x_v$ of the polyhedron in question. The following features of a Gale transform of a polyhedron should be noted:

1. Gale transforms $\bar{x}_j$ and $\bar{x}_k$ of two or more vertices of a polyhedron may lead to the same point (i.e., the same v–4 coordinates) in (v–4)-dimensional space ($\Re^{v-4}$). In other words, some points of a Gale transform may have a multiplicity greater than one so that the Gale transform of a polyhedron in such cases contains fewer *distinct* points than the polyhedron has vertices.
2. The Gale transform depends upon the location of the origin in the coordinate system. Therefore, infinitely many Gale transforms are possible for a given polyhedron. Geometrically a Gale transform of a polyhedron corresponds to a projection of the v vertices of a (v–1)-dimensional simplex (i.e., the higher dimension "analogue" of the

tetrahedron in three dimensions) into a (v–4)-dimensional hyperplane.[45] Since infinitely many such projections are possible, the Gale transform for a given polyhedron is *not* unique.

In practice, it is easier to work with *Gale diagrams* corresponding to Gale transforms of interest. Consider a Gale transform of a (three-dimensional)

polyhedron having v vertices $\bar{x}_1,\ldots\bar{x}_v$ as defined above. The corresponding Gale diagram $\hat{x}_1,\ldots,\hat{x}_v$ is defined by the following relationships:

$$\hat{x}_i = 0 \text{ if } \bar{x}_i = 0 \qquad (12\text{–}8a)$$

$$\hat{x}_i = \frac{x_i}{\|\bar{x}_i\|} \text{ if } \bar{x}_i \neq 0 \qquad (12\text{-}8b)$$

In Equation 12–8b, $\|\bar{x}_i\|$ is the length (i.e.,$\sqrt{a^2{}_{i,1} + a^2{}_{i,2} + \ldots + a^2{}_{i,v-4}}$) of the vector $\bar{x}_i$. If v–4 = 1 (i.e., v = 5), Gale diagrams can only contain the points of the straight line 0, 1, and –1 of varying multiplicities m_0, m_1, and m_{-1}, respectively, where $m_0 \geq 0$, $m_1 \geq 2$, and $m_{-1} \geq 2$. If v–4 = 2 (i.e., v = 6), Gale diagrams can only contain the center and circumference of the unit circle. These two types of Gale diagrams (Figure 12–8) are of interest for the study of polyhedral isomerizations, since they represent significant structural simplifications of the corresponding polyhedra.

The following properties of Gale diagrams corresponding to three-dimensional polyhedra are of interest, since they impose important restrictions on configurations of points which can be Gale diagrams:

1. Any (v–5)-dimensional plane passing through the central point of the Gale diagram bisects the space of the Gale diagram into two halfspaces. Each such halfspace must contain at least two vertices (or one vertex of multiplicity 2) of the Gale diagram, *not* including any vertices actually in the bisecting plane or hyperplane. Such a halfspace is called an *open* halfspace. Violation of this condition corresponds to a polyhedron with the impossible property of at least one pair of vertices *not* connected by an edge which is closer in three-dimensional space than another pair of vertices which is connected by an edge.
2. The set of vertices of a polyhedron *not* forming a given face or edge of the polyhedron is called a *coface* of the polyhedron. The regular octahedron is unusual, since all of its faces are also cofaces corresponding to other faces. The interior of a figure formed by connecting the vertices of a Gale diagram corresponding to a coface must contain the central point.

[45] P. McMullen and G. C. Shephard, *Mathematika*, **15**, 223, 1968.

3. The central point is a vertex of a Gale diagram if and only if the corresponding polyhedron is a pyramid. The central vertex of such a Gale diagram corresponds to the apex of a pyramid, which is the coface corresponding to the base of the pyramid.

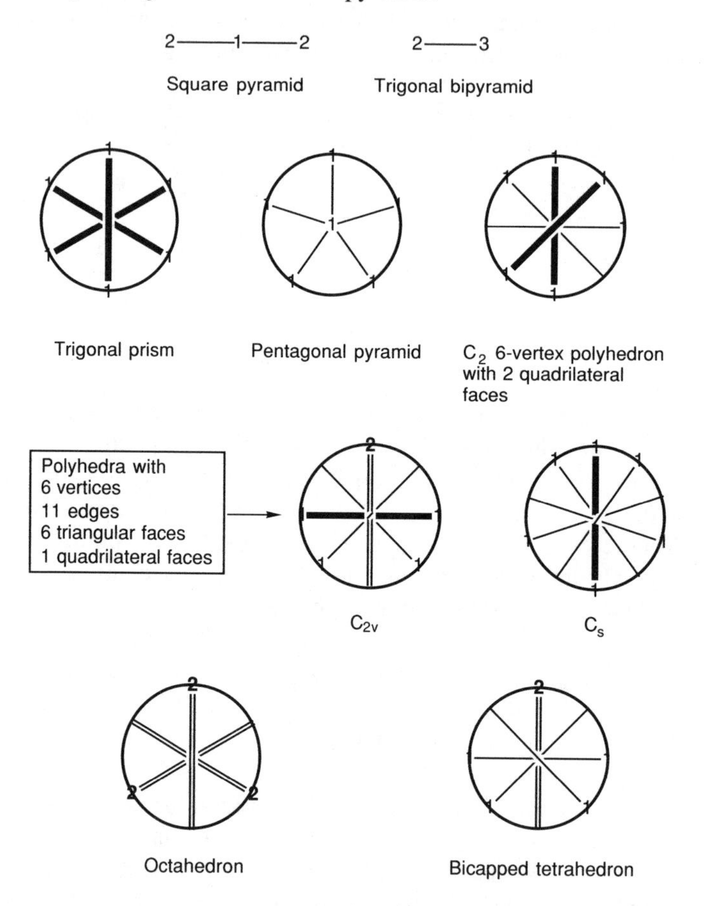

FIGURE 12–8. Gale diagrams for the two combinatorially distinct five-vertex polyhedra and the seven combinatorially distinct six-vertex polyhedra. The multiplicities of the vertices are labelled, double lines are used for diameters leading to vertices of multiplicity 2, and bold lines are used for balanced diameters. The number of balanced diameters (i.e., number of bold diameters) is equal to the number of quadrilateral faces.

Nonplanar isomerizations of five- and six-vertex polyhedra correspond to allowed vertex motions in the corresponding Gale diagrams in Figure 12–8. In this context, an *allowed vertex motion* of a Gale diagram is the motion of one or more vertices which converts the Gale diagram of a polyhedron into that of another polyhedron with the same number of vertices without ever passing through an impossible Gale diagram, such as one with an open halfspace containing only one vertex of unit multiplicity. Since two polyhedra are combinatorially equivalent if and only if their Gale diagrams are isomorphic, such allowed vertex motions of Gale diagrams are faithful representations of all possible non-planar polyhedral isomerizations.

The application of Gale diagrams to the study of isomerizations of five-vertex polyhedra is nearly trivial but provides a useful illustration of this method. The only possible five-vertex polyhedra are the square pyramid and trigonal bipyramid. Their Gale diagrams (Figure 12–8) are the only two possible one-dimensional five-vertex Gale diagrams which have the required two vertices in each open halfspace (i.e., $m_1 \geq 2$ and $m_{-1} \geq 2$). The only allowed vertex motion in a Gale diagram of a trigonal bipyramid involves motion of one point from the vertex of multiplicity 3 through the center point to the vertex originally of multiplicity 2 (Figure 12–9). This interchanges the vertices of multiplicities 2 and 3 and leads to an equivalent Gale diagram corresponding to an isomeric trigonal bipyramid. The motion through the center point of the Gale diagram corresponds to the generation of a square pyramid intermediate in the non-planar degenerate isomerization of a trigonal bipyramid. This, of course, is the Berry pseudorotation process, which is the prototypical dsd process (see Section 12.3). The choice of three points to move away from the vertex of multiplicity 3 in the Gale diagram of a trigonal bipyramid corresponds to the presence of three degenerate edges in a trigonal bipyramid. This analysis of the Gale diagrams of the two possible five-vertex polyhedra shows clearly that the only possible nonplanar isomerizations of five-vertex polyhedra can be represented as successive dsd processes corresponding to successive Berry pseudorotations.

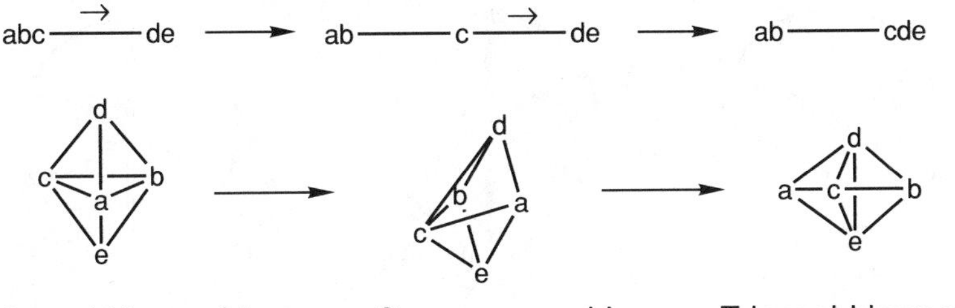

FIGURE 12–9. The Gale diagrams corresponding to the single dsd degenerate isomerization of a trigonal bipyramid through a square pyramid intermediate (Berry pseudorotation). Vertices in Figures 12–9 and 12–10 are labelled with lower case letters, and the motions in the Gale diagrams are indicated by arrows. In this isomerization, the edge connecting vertices a and b is broken, and a new edge between vertices d and e is formed.

The Gale diagrams of six-vertex polyhedra can be visualized most clearly if all of the diameters containing vertices are drawn, as is done in Figure 12–8. Some Gale diagrams of six-vertex polyhedra have diameters with vertices of unit multiplicity at *each* end; such diameters may be called *balanced diameters* and are indicated by bold lines in Figure 12–8. The two vertices of a balanced diameter in the Gale diagram of a six-vertex polyhedron form an edge which is a coface corresponding to a quadrilateral face. Gale diagrams drawn to maximize the multiplicities of the vertices and the numbers of balanced diameters consistent with the polyhedral topology are called *standard* Gale diagrams. The Gale diagrams depicted in Figure 12–8 are the standard Gale diagrams for the six-vertex polyhedra in question. The number of balanced diameters in a standard Gale diagram of a six-vertex polyhedron is equal to the number of quadrilateral faces of the polyhedron. The pentagonal pyramid is the only six-vertex polyhedron for which the center of the circle is a vertex of the corresponding standard Gale diagram.

Polyhedral isomerizations in six-vertex polyhedra may be represented by allowed motions of the vertices of their Gale diagrams along the circumference of the unit circle or through the center, in the case of polyhedral isomerizations involving a pentagonal pyramid intermediate. However, vertex motions are not allowed if at any time they generate one or more *forbidden diameters* containing three (or more) vertices.

Using these techniques, all non-planar degenerate isomerizations involving six-vertex polyhedra can be decomposed into sequences of eight fundamental processes,[46] namely two processes through pentagonal pyramid intermediates, five processes which are variations of single diamond-square-diamond processes (Section 12.2), and the triple dsd degenerate isomerization of an octahedron through a trigonal prism intermediate on which the Bailar[31] and Ray and Dutt[32] twists of $M(bidentate)_3$ complexes are based. The Gale diagrams for this last process are depicted in Figure 12–10.

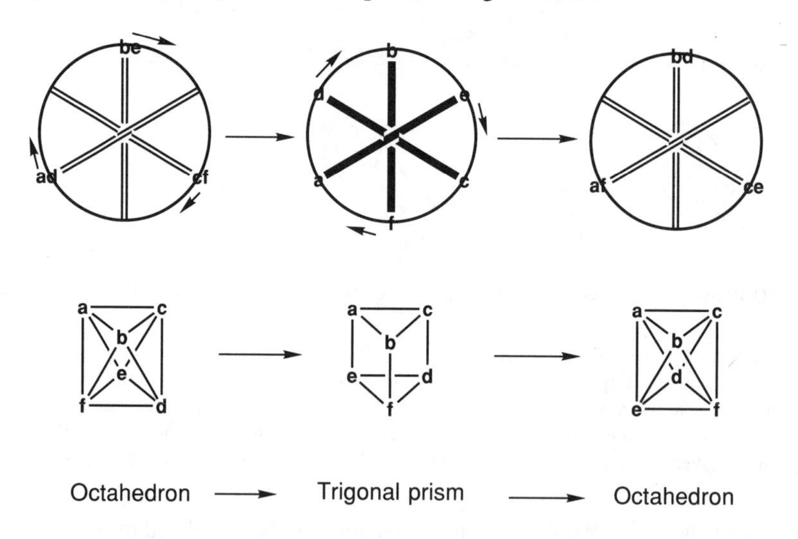

FIGURE 12–10. Standard Gale diagrams for the triple dsd degenerate isomerization of an octahedron through a trigonal prism intermediate (Bailar or Ray and Dutt Twist).

[46]R. B. King, *Theor. Chim. Acta*, **64**, 439, 1984.

INDEX

A

Abelian groups, 20, 30
Ab initio methods, 71, 73
Abstract definition, 30
Accidental orthogonality, 161
Achiral groups, 33, 194
Acidic melts, 125
Actinide, 165
Adjacency matrix, 6, 49, 50, 54, 71
Adjustment factor X, 85
Alkali metals, 125, 130, 131, 134, 136–137, 181, see also specific types
Alky groups, 83
Alkylarsinidene, 83, 96
Alkylphosphinidene, 83, 91, 92, 96
Alkylphosphinidine, 96
Allene, 34, 179
Allotropic forms, 174
Alternating groups, 30, 171
Aluminum, 178–181, 184
Aluminum icosahedra, 173
Angular component of wave functions, 36
Angular coordinates, 36, 67
Anisotropic conductivity, 152
Anodal orbital aromaticity, 54
Anthracene, 87, 105, 107–109, 151, 162
Antibonding orbitals, 47, 50, 53, 58, 65
  core, 78, 117
  distribution of, 74
  missing, 85
  pure, 79
  surface, 79
Antiferromagnetism, 159, 160, 170
Antimony, 43, 98, 103, 125, 126, 129
Antipodal interactions, 78
Antipodal pairs, 74, 76, 77
Antipodal vertices, 74, 102
Antiprisms, see also specific types
  bicapped square, 102, 128, 196
  capped square, see Capped square antiprisms
  oblate, 174
  pentagonal, 102–103
  square, see Square antiprisms
  tetragonal, 100–101
  trigonal, 34, 174
Antisymmetric irreducible representations, 25, 26
Apical vertices, 62, 88
Apparent electron deficiencies, 133
Apparent skeletal electrons, 84, 86–89, 91, 92
  bicapped tetrahedra and, 93
  capped octahedra and, 93
  pentagonal antiprisms and, 103
  tricapped trigonal prisms and, 96
Arachno structures, 61, 99, see also specific types
  borane, 86
  polyhedral, 62, 63, 66, 101, 102
Armstrong-Perkins-Stewart computations, 79
Aromatic hydrocarbons, 150, see also specific types
  planar, 34, 47, 54, 71, 163
  polycyclic, 105, 150
  polygonal, 71
Aromaticity, 47, 53, 54
  benzenoid-type, 179
  binodal, 161–168
  three-dimensional, 55, 126, 127
  two-dimensional, 127
  uninodal, 165
Aromatic stabilization, 55, 59, 60
Aromatic systems, 165, see also specific types
Arsenic, 126
Associative law of multiplication, 19
Atomic orbitals, 35–37, 47, 49, 71
  energy of, 48, 72
  hybrid, see Hybrid orbitals
  irreducible representations of, 40
  linear combination of, 47, 79
  metal valence, 38
  orthogonal nature of, 37
  overlap between, 37
  properties of, 37
  shape of, 37
  spherically symmetric, 37
  valence, 38
  vertex, 139
Atoms, see also specific types
  axial, 90
  basal, 184
  border, 62, 64, 66
  equatorial, 90
  heavy, 50, 82
  interior, 62, 65, 66
  interstitial, 97–105, see Interstitial atoms
  light, 50, 86, 98
  oxygen, 159–163, 165
  single interior vertex, 65
  vertex, see Vertex atoms
Audier model, 183, 184, 186
Automorphisms, 76, 172, 202

Axes, 21, 25, see also specific types
- in perpendicular plane, 22
- reference, 22
- rhombohedral, 174
- rotation, 10, 17, 18, 21, 25, 32

Axial atoms, 90
Axial positions, 198
Axial vertices, 22, 89, 196
Azimuthal quantum number, 36

B

Bardeen-Cooper-Schrieffer theory, 149
Bare metal clusters, 125–130
Bare phosphorus vertices, 83
Basal atoms, 184
Basal positions, 114
Basal vertices, 62
Base faces, 13
Basis sets, 48
Belts of bipyramids, 85
Benzene, 51–55, 62, 64
- aromatic stabilization of, 55
- dehydrogenation of cyclohexane to, 100
- external orbitals and, 82
- fusion of, 107, 109
- globally delocalized, 92
- infinite solid state structures and, 152
- metal oxides and, 163
- NMR spectra of, 92

Benzenoid hydrocarbons, 105, 162
Benzenoid-type aromaticity, 179
Berry pseudorotation, 196, 209
Bicapped octahedra, 95
Bicapped square antiprisms, 102, 128, 196
Bicapped tetrahedra, 16, 86, 87, 92–93
Bicapped trigonal prisms, 43, 45, 46
Binodal orbital aromaticity, 161–168
Binodal orbitals, 67
Bipartite graphs, 4
Biphenyl, 106, 109
Bipyramids, 85, see also specific types
- belts of, 85
- hexagonal, 39, 45, 46, 201, 202
- pentagonal, see Pentagonal bipyramids
- tetragonal, 85
- trigonal, see Trigonal bipyramids

Bisdisphenoid, 44, 45, 101, 198
- irreducible representations of, 43
- microscopic models of, 196
- orbital representations for, 73
- polyhedral dynamics and, 201, 202
- post-transition metal clusters and, 131
- symmetry of, 201

Bisdisphenoid enantiomer pairs, 203
Bismuth, 125, 126, 129, 170
Bismuth copper oxide superconductors, 168
Bisphenoid (allene) skeleton, 34
Block-factored matrix, 23
Bonding, chemical, see Chemical bonding
Bonding orbitals, 47, 50, 53, 66
- core, 56
- icosahedra and, 184
- irreducible representations of, 117
- metal oxides and, 165
- post-transition metal clusters and, 137
- skeletal, 69, 69.90

Boranes, 56, see also specific types
- *arachno,* 86
- delocalization of, see Delocalization
- deltahedral, see Deltahedral boranes
- dianions of, 51
- global delocalization of, 53–61
- halides and, 113
- icosahedral, see Icosahedral boranes
- *nido,* 86
- octahedral, 73–80
- polyhedral, 37, 51, 52, 125
- polyhedral dynamics and, 196

Border atoms, 62, 64, 66
Borides, 156, 178
- boron icosahedra in, 173–179
- boron-rich, 178, 181
- lanthanide rhodium, 153, 169
- metal, 173–179, 181
- ternary lanthanide rhodium, 153

Boron, 50, 51, 83, 86
- chemistry of, 65
- clusters of, 127
- elemental, 173–179, 181, 186
- equatorial, 175
- icosahedral, 176
- infinite solid state structures and, 153, 154
- interstitial, 175–179
- isolated, 179
- polyhedral, 130
- quasicrystals and, 181
- rhombohedral, 174–176, 178, 183
- tetracoordinate, 177
- vertices of, 86, 131

Boron hydride, 61
Boron icosahedra, 173–179
Boron-rich borides, 178, 181
Bragg diffraction, 180
Bulk metal structures, 105, 139–145
Butterflies, 63, 127

C

Capped deltahedra, 86, 129
Capped octahedra, 12, 44, 86, 87, 197

electron counting and, 93–96
irreducible representations of, 43
polyhedral dynamics and, 196
Capped polyhedra, 12, 87
Capped square antiprisms, 43, 46, 102, 114, 127, 129–130
Capped tetrahedra, 86
Capped trigonal prisms, 43, 44
Capping, 12, 85
Carbide, 101
Carbon, 50, 51, 83, 98, 99
antiprisms and, 100
clusters of, 189
elemental, 186–191
hexagons of, 105, 107
icosahedra and, 179
interstitial atoms and, 101
planar, 187
vertices of, 86
Carbonyl groups, 125
Carboranes, 51, 82, see also specific types
deltahedral, 53
global delocalization of, 53–61
halides and, 113
polyhedral, 51, 52, 125
polyhedral dynamics and, 196
post-transition metal clusters and, 127, 130, 135
trigonal bipyramidal, 197
Centered metal clusters, 116–123
Central icosahedra, 185
Central points, 32
Chalcogenides, 148
Chalcogens, 152
Characteristic polynomials, 6, 9, 10
Character of matrices, 24
Character tables, 24, 26, 173
Chemical bonding, 38
core, see Core bonding
delocalized, see Delocalized chemical bonding
edge-localized, 86, 93, 100, 134, 149
external, 176
internal, 176
localized, 47, 52, 86, 134, 149
multicenter, 131, 133, 139, 147
skeletal, see Skeletal bonding
surface, see Surface bonding
topology of, see Chemical bonding topology
two-center, 47
Chemical bonding topology, 53, 54, 56, 71, 113, 139, 142
of icosahedra, 176
of superconductors, 153, 158
Chemical dynamics, 2
Chemical species, 2
Chevrel phases, 148, 149, 153, 169
Chirality, 32, 33
Chiral polyhedra, 194
Chlorine, 116
Chromium, 143, 145, 147
Circles, 210
Circuit (cyclic) graphs, 3, 54, 55
Circuits, 5–8, see also specific types
Cis interactions, 57, 58, 74
Closed deltahedra, 62, 63, 65, 86, 102
Closure condition, 198
Cluster skeletal bonding, 83–85
Cluster valence molecular orbitals (CVMO), 84–85
Cobalt, 100
Cofaces, 207
Coherence length, 149
Combinatorially distinct polyhedra, 15
Combinatorially equivalent, 13, 208
Commutative groups, 20
Complement, 3
Complete core bonding topology, 76
Complete graphs, 3–5, 56, 59, 64, 66
anthracene and, 108
bipartite, 4
early transition metal clusters and, 120
fused octahedra and, 107
icosahedrally weighted, 76, 77
octahedra and, 108
symmetric groups of, 79
tensor surface harmonic theory and, 67
tetrahedra and, 196
topology of, 90
Complete topology, 58, 59
Components, 6
Condensed-phase metal clusters, 130
Conducting skeleton, 148, 156, 158, 168, 169
Conduction-electron wave functions, 149
Conductivity, 145, 152, 170
Conjugacy classes, 20, 25, 29, 31, 171–173
Conjugate operations, 20
Connected graphs, 5, 6
Connectivity, 1, 198, 200
Cooper electrons, 147
Coordination complexes, 194, 196
Coordination numbers, 38, 42–46
Coordination polyhedra, 38–46
coordination numbers and, 42–46
forbidden, 45
irreducible representations of, 41, 43
spd-forbidden, 45
Copper, 181, 183, 184
Copper complexes, 161
Copper-copper interactions, 159

Copper oxides, 161, 168–170
Copper oxide superconductors, 146, 159
Core bonding, 108
  fused octahedra and, 107
  of icosahedra, 173, 176
  multicenter, 90, 109
  of octahedra, 141
  post-transition metal clusters and, 135
  topology of, 58, 76
Core bonding electron pairs, 59
Core bonding graphs, 56
Core-external orbial mixing, 77
Core orbitals, 57, 76, 78
  antibonding, 78, 117
  bonding, 56
  doubly degenerate, 74
  energy of, 75, 77, 78
  icosahedral, 77
  octahedral, 75
  principal, 57, 72, 78
  pure, 76
  symmetric, 68
Core-surface mixing, 68, 72, 75, 76, 78, 79
Coulomb integrals, 71
Coupled metals, 160, see also specific types
Covalent divide, 144
Critical current, 146
Critical fields, 148, 149, 153
Critical temperatures, 146, 149, 153
Cross-polytope, 202
Crystallography, 180, 181
Cubanes, 52
Cube enantiomers, 203
Cubes, 13, 39, 46, 52, 85
  centered, 120
  early transition metal clusters and, 114, 120, 123
  electron counting and, 97
  faces of, 183
  infinite solid state structures and, 148, 153, 156, 157
  irreducible representations of, 43
  omnicapped, 97
  polyhedral dynamics and, 201, 202
  post-transition metal clusters and, 137
  reference, 157
  symmetry of, 201
Cuboctahedra, 98, 99, 104–105, 163, 164, 165
Curvature, 187
CVMO, see Cluster valence molecular orbitals
Cycle index, 29
Cycle structures, 29, 172
Cyclic (circuit) graphs, 3, 54, 55
Cyclic subgroups, 30
Cycloalkanes, 163
Cyclobutadiene, 64, 127
Cycloheptatrienyl, 82
Cyclohexane, 92, 100, 163
Cyclohexatriene, 55
Cyclopentadienide anion, 129
Cyclopentadienyl, 62, 64, 82, 125
Cyclopropenyl cations, 129
Cylindrical geometry, 119

D

Dawson structure, 165
Decarbonylation, 100
Degeneracy, 58
Degeneracy-weighted average distance, 72
Degeneracy-weighted mean of energy parmaeters, 79
Degenerate edges, 196
Degenerate isomerization, 197, 198, 199, 210
Degenerate orbitals, 64, 72
Degenerate polyhedral isomerization, 193
Dehydrogenation, 100
Delocalization, 47–69, 71–80, 85
  defined, 51
  of deltahedra, 84, 85
  electron rich polyhedra and, 61–66
  global, see Global delocalization
  graph theory and, 47–50
  Huckel theory and, 47–50
  icosahedra and, 187
  of icosahedral boranes, 73–80
  infinite, 140, 153
  metal oxides and, 161–168
  molecular orbital parameters in, 71
  of octahedral boranes, 73–80
  porous, 149, 153
  tensor surface harmonic theory and, 67–69
  topological analysis of computational results on, 71–72
  vertex atoms and, 50–52
Delocalized chemical bonding, 3, 47, 49, 52, 55, 59, 89
  in polyhedral cavities, 184
Deltahedra, 46, 52, 55, 61, 67, 86, 190
  capped, 86, 129
  closed, 62, 63, 65, 86, 102
  core bond of, 67
  degenerate edges in, 196
  delocalized, 84, 85
  diamond-square-diamond processes in, 195–198

edge-coalesced, 135
eleven-vertex, 197
gallium, 131
globally delocalized, 84, 85, 131
interstitial atoms and, 98
nine-vertex, 107
orbital representations for, 73
point groups of, 68
post-transition metal clusters and, 130, 131, 135
pseudorotation of, 197
punctures of surface of, 61
rearrangements of, 195
structures of, 197
three-dimensional, 63
topological homeomorphism of, 67
Deltahedral boranes, 47, 51, 53–57, 197
anions of, 72, 82
energy parameters and, 72
internal orbitals in, 168
topological methods for, 71
Deltahedral carboranes, 53
Deltahedral clusters, 61
Deltahedral graphs, 56, 69
Deltahedral topology, 56, 59
Desargues-Levy graphs, 199
Determinantal equations, 9, 49
Diamond pairs, 156
Diamonds, 186, 187
Diamond-square processes, 98, 99, 101, 105
Diamond-square-diamond processes, 98, 195–198, 209, 210
Differential equations, 35, 67
Diffraction, 90, 91, 125, 126, 180
Direct product groups, 20, 25–26, 27, 172
Disconnected graphs, 5, 10, 53, 59
Discrete energy, 139
Discrete icosahedra, 174
Distortion, 120, 127
o-Divinylbenzene, 7
Dodecahedra, 44, 52, 85, 182, 188, 189, 198
Dodecahedranes, 52
D orbitals, 81
Double decarbonylation, 100
Double diamond-square process, 101
Double primes, 25
Doubly degenerate core molecular orbitals, 74
Doughnut holes, 187
Doughnuts, 118
Dsd, see Diamond-square diamond
Dualization, 12, 13, 190
Dual polyhedra, 13, 182
Dual processes, 86, see also specific types

E

Early transition metal clusters, 142, see also specific types
gold, 118–123
halide, 113–117
Early transition metal polyoxometalates, 161–168
Edge-coalesced deltahedra, 135
Edge-coalesced icosahedra, 55–56, 135, 197, 198
Edge-localized chemical bonding, 86, 93, 100, 134, 149
Edge-localized clusters, 88
Edge-localized octahedra, 113–117, 149
Edge-localized polygons, 51
Edge-localized polyhedra, 51, 134
Edge-localized skeltal bonding, 100
Edge-localized tetrahedra, 153
Edges, 1–3, 8, 198
degenerate, 196
equatorial-equatorial, 198
face relationship with, 11
of icosahedra, 76, 77
incident, 87
midpoints of, 202
number of, 188
of octahedra, 57, 74
switching, 195, 196
vertex relationship with, 11
vertical, 196
Edge-sharing fusion, 105
Edge-sharing naphthalene, 106–107
Eigenvalues, 6, 50, 54, 55, 56, 61, 68
of adjacency matrix, 71
doubly degenerate core molecular orbitals and, 74
of one-vertex no-edge graphs, 64
positive, 69, 121, 122, 123, 137
principal, 56, 165
zero, 137
Eight-coordination, 39
Electrical conductivity, 145
Electrical resistance, 146
Electron counting, 83, 84–85
icosahedra and, 176
infinite solid state structures and, 139
in metal carbonyl clusters, 84–85, 87–97, 109
bicapped tetrhedra and, 92–93
capped octahedra and, 93–96
cubes and, 97

icosahedra and, 96–97
octahedra and, 90–96
omnicapped cubes and, 97
tetrahedra and, 87, 92–93
tricapped trigonal prisms and, 96
trigonal bipyramids and, 88–90
trigonal prisms and, 96
total, 84
Electron-electron repulsion, 145
Electron-poor systems, 85–87, 93, 135
Electron poverty, 86
Electron-precise bonding models, 3
Electron-rich polyhedra, 61–66, 85, 86
Electrons, see also specific types
cooper, 147
counting of, see Electron counting
distribution of, 3
extra, 85
interstitial, 145
latent, 117
mobile, 147
phopshorus valence, 83
skeletal, see Skeletal electrons
transport of, 168
valence, 83, 126
Electron spin resonance (ESR), 167
Electropositive metals, 178, 179, 184, see also specific types
Elemental boron, 173–179, 181, 186
Elemental carbon cages, 186–191
Elongation, 85
Enantiomers, 194, 198, 199, 202, 203
Energy, 50, 71, 72, 76
atomic orbital, 48, 72
core orbital, 75, 77, 78
degeneracy-weighted means and, 79
discrete, 139
gap in, 160
Huckel orbital, 50
of icosahedra, 78
of icosahedral core orbitals, 77
of interactions, 48
levels of, 9, 67
matrices of, 49
molecular orbital, 50, 70–72, 78
of octahedral core orbitals, 75
potential, 35
semiconducting, 141
zero point of, 71
Energy units, 54, 72, 76
Equatorial atoms, 90
Equatorial boron, 175
Equatorial-equatorial edges, 198
Equatorial vertices, 22, 89, 174, 196, 198
Equator of icosahedra, 174
Equilateral triangles, 34, 129, 171
ESR, see Electron spin resonance
Eulerian graphs, 5–6
Eulerian trail, 6
Euler's relationship, 11, 84, 188, 189
Even-order improper rotation axis, 21
Even permutations, 28, 171
Exchange integrals, 160
External bonding, 176
External faces, 142
External orbitals, 50, 82, 83, 89, 116
of bare metal atoms, 126
early transition metal clusters and, 119
icosahedra and, 174, 182, 185
infinite solid state structures and, 142
nonbonding, 82
post-transition metal clusters and, 136
External-surface mixing, 72
Extra electron pairs, 85

F

Face bonds, 114, 115
Face-fused octahedra, 152
Face-linked octahedra, 152
Face-localized octahedra, 113–117
Faces, 11, 188
base, 13
of cubes, 183
edge relationship with, 11
external, 142
hexagonal, 187, 189
non-triangular, 99
pentagonal, 85, 183, 185, 186, 188, 200
pivot, 195
pseudohexagonal, 199
quadrilateral, 11, 98, 101, 195, 210
square, 14
totality of, 11
triangular, see Triangular faces
Face-sharing fusion, 151
Face-sharing naphthalene, 107
Face-sharing octahedra, 108
False metal-metal bonds, 85
Fermi velocity, 149
Ferromagnetism, 159, 160, 161
Finite groups, 18
Five-dimensional representations, 24
Five-fold degeneracy, 58
Fixed points, 29
Fluxional processes, 193, 195, 197
Forbidden polyhedra, 26, 45, 196, 202
Four-dimensional representations, 24
Fourier transforms, 180

Framework groups, 32–34
Fullerenes, 186–191
Fused octahedra, 105–110, 142, 150
Fused polyhedra, 108
Fused tetrahedra, 86, 92
Fusion, 109
  benzene, 107, 109
  edge-sharing, 105
  face-sharing, 151
  of icosahedra, 175
  of macrocuboctahedra, 164
  of metal cluster octahedra, 139–145
  of octahedra, 139–145, 149

G

Gadolinium, 140, 141
Gale diagrams and transformations, 204–210
Gallium, 126, 130–135, 181
Gallium-gallium bond, 134
Gallium icosahedra, 131
Gas phase post-transition metal clusters, 128–130
Gaussian computations, 79
Gaussian functions, 79
Gaussian orbitals, 79, 80
Generators, 30
Geodesic domes, 187
Germanium, 103, 126, 127
Gimarc-Ott analysis, 197
Global delocalization, 51–61, 88, 91
  aromatic stabilization and, 55, 60
  of boranes, 53–61
  of carboranes, 53–61
  of clusters, 92
  of deltahedra, 84, 85, 131
  deltahedra and, 57, 59, 61
  of hydrocarbons, 53–61
  internal orbitals and, 54
  of octahedra, 113–117
  of octahedral cavities, 152
  octahedral specta and, 59
  of polygons, 51, 60
  of polyhedra, 51, 60, 135
  skeletal bonding and, 60
  of trigonal prisms, 96
Gold, 118–123, 136
Goldberg polyhedra, 189, 190
Graphite, 179, 186, 187
Graphs, see also specific types
  bipartite, 4
  complete, see Complete graphs
  connected, 5, 6
  core bonding, 56
  cyclic, 3, 54, 55
  defined, 1
  deltahedral, 56, 69
  Desargues-Levy, 199
  disconnected, 5, 10, 53, 59
  Eulerian, 5–6
  Hamiltonian, 5–6
  homeomorphic, 5
  hyperoctahedral, 201
  isopectral, 7
  line, 4
  non-planar, 5
  numberings of, 6
  octahedral, 59
  odd, 199
  Petersen's, 199
  planar, 4–5, 15
  polygonal, 3
  regular, 7
  Sachs (sesquivalent subgraphs), 7, 8
  spectra of, 6–10, 122
  subgraphs of, see Subgraphs
Graph theory, 3–6
  defined, 1
  delocalization and, 47–50, 53
  Huckel theory compared to, 121
Great orthogonality theorem, 41
Group representations, 23–25, 33, see also specific types
Groups, see also specific types
  Abelian, 20, 30
  achiral, 33, 194
  alternating, 30, 171
  automorphism, 76, 202
  commutative, 20
  direct product, 20, 25–27, 172
  finite, 18
  framework, 32–34
  icosahedral, 21, 31
  linear, 21–22
  metacyclic, 30
  monovalent, 83
  octahedral, 21
  permutation, see Permutation groups
  point, see Point groups
  polyhedral, 21–22
  representation of, 23–25
  simple, 20
  soluble, 20
  symmetric, 27, 28
  symmetry, 27, 28, 33, 79, 201
  symmetry point, see Symmetry point groups
  tetrahedral, 21, 31
  transitive, 30
  wreath product, 201
Group theory, 58, 67

H

Half icosahedra, 175
Halfspaces, 34, 86, 207, 208
Halides, 125
Hall coefficients, 170
Halogen, 50, 83, 113–115, 140, 142
Halometalate salts, 125
Hamiltonian circuit, 6
Hamiltonian graphs, 5–6
Harmonics
  spherical, 35, 36, 67
  surface, 37, 56, 67–69, 115, 117
  tensor surface, 115, 117
  vector, 68
Heat capacity, 145
Heavy atoms, 50, 82
Hellmann-Feynmann theorem, 145
Heptagonal pyramids, 201
Heptagons, 34
Heteropolyoxometalates, 161
Hexagonal bipyramids, 39, 45, 46, 201, 202
Hexagonal faces, 187, 189
Hexagonal pyramids, 64
Hexagonal rings, 179
Hexagons, 3, 87, 187, 189
  carbon, 105, 107
  infinite solid state structures and, 141, 143
  planar, 34, 150
  polyhedral dynamics and, 202
Hidden two-center two-electron bonds, 109, 110
High critical temperature copper oxide superconductors, 159
Higher-order rotation axis, 21
Hoffmann-Lipscomb computations, 73, 78, 79
Homeomorphism, 5, 67
Homonuclear trigonal bipyramids, 90
Horizontal plane of symmetry, 22, 25
Horizontal reflection planes, 45
Huckel methods, 48, 71, 73, 119, 165
Huckel molecular orbitals, 49, 50
Huckel parameters, 50
Huckel rule, 189
Huckel theory, 47–50, 121
Hume-Rothery rule, 105
Hybridization, 62
Hybrid orbitals, 38–46
  coordination numbers and, 42–46
  irreducible representations of, 43
  tetrahedrally, 83
Hydrocarbons, see also specific types
  aromatic, see Aromatic hydrocarbons
  benzenoid, 105, 162
  delocalization of, see Delocalization
  global delocalization of, 53–61
  planar, see Planar hydrocarbons
  polycyclic, 105, 150
  polygonal, 51, 53, 62, 65, 71, 81
Hydrogen, 50, 81, 82, 83, 119, 142
Hydrogenation, 92
Hyperoctahedra, 201, 202
Hyperoctahedral reduction, 202
Hyperoctahedral restriction, 201, 202
Hypersurfaces, 180, see also specific types
Hypervalent alkylphosphinidine groups, 96
Hypervalent vertices, 83, 91, 96
Hypho polyhedra, 66

I

Icosahedra, 20, 72, 77, 102, 171–191
  aluminum, 173
  boron, 173–179
  central, 185
  chemical bonding topology of, 176
  core bonding of, 173, 176
  discrete, 174
  early transition metal clusters and, 120
  edge-coalesced, 55–56, 135, 197, 198
  edges of, 76, 77
  electron counting and, 96–97
  electron-rich, 63
  elemental carbon cages and, 186–191
  energy parameters for, 78
  equator of, 174
  fullerenes and, 186–191
  fusion of, 175
  gallium, 131
  gold, 123
  half, 175
  of icosahedra, 179
  infinite solid state structures and, 156
  interstitial atoms and, 103–104
  Mackay, 185, 186
  orbital representations for, 73
  oxygen atoms and, 165
  polyhedral dynamics and, 197
  post-transition metal clusters and, 133, 134
  quasicrystals of, 173, 180–186
  regular, 62, 79
  skeletal bonding of, 175
  surface bonding of, 173, 175
  surface orbital energy parameters for, 78
  symmetry of, 171–173, 182

symmetry point groups of, 171
truncated, 175, 182, 183, 187, 189
Icosahedral boranes, 173
delocalization of, 73–80
Icosahedral boron, 176
Icosahedral cavities, 105, 175
Icosahedral groups, 21, 31
Icosahedrally weighted complete graphs, 76, 77
Icosidodecahedra, 185
Identity, 17, 18, 19, 22, 26
Identity permutations, 27, 28, 193
Improper rotations, 17, 25, 32–33, 173
Incident edges, 87
Indexed variables, 29
Indium, 126, 130–135
Infinite delocalization, 140, 153
Infinite metal clusters, 140
Infinite order rotation axis, 21
Infinite solid state structures, 139–158, 173
anthracene and, 151
fusion of metal clusters and, 139–145
naphthalene and, 150–151
superconductors and, see Superconductors
Insolubility, 144
Integrals, 48, 71, 160, see also specific types
Intericosahedral bonding, 175
Interior atoms, 62, 65, 66
Intermediate polyhedra, 193, 197
Internal bonding, 176
Internal orbitals, 50, 52, 67, 107
capped octahedra and, 94
cluster skeletal bonding and, 83
early transition metal clusters and, 120
extra, 87, 92
globally delocalized octahedra and, 113
infinite solid state structures and, 139–143, 150, 154
live, 109
metal oxides and, 168
normal, 89, 93
perinaphthene and, 108
post-transition metal clusters and, 126
rhodium and, 90
total, 141
twin, 53, 62, 65, 126
unique, 59, 61, 62, 65, 66, 72, 126
Internal species, see Chemical species
Interoctahedral metal-metal interactions, 148
Interpenetrating loops, 162
Interpolyhedral skeletal bonding, 50
Interstitial atoms, 97–105, see also specific types
boron, 175–179
in centered metal clusters, 117
cuboctahedra and, 104–105
deltahedra and, 98
icosahedra and, 103–104
nitrogen, 116
octahedra and, 99
single, 116
trigonal prisms and, 100
Interstitial electrons, 145
Intervertex repulsion, 193
Intrapolyhedral skeletal bonding, 50, 131
Inverses, 19
Inversion, 18, 26
Inversion centers, 10, 18, 25, 27, 45
Involution, 26, 45
Ionic bare post-transition metal clusters, 125–130
Ionic divide, 144
Iron, 91
Iron alloys, 144
Irred reps, see Irreducible representations
Irreducible representations, 23, 24, 39, 40
antisymmetric, 25, 26
of atomic orbitals, 40
of bonding skeletal orbitals, 117
of coordination polyhedra, 41, 43
delocalization and, 72
five-dimensional, 24
four-dimensional, 24
of hybrid orbitals, 43
odd, 45
one-dimensional, 24–25
symmetric, 25, 26, 56
of symmetry point groups, 24–25
three-dimensional, 24
two-dimensional, 24
ungerade, 40
Isocyanide ligands, 82
Isocyanides, 83, 119, 121
Isolated boron, 179
Isolated tetrahedra, 86
Isolobality, 82
Isomer count, 194, 198
Isomerizations, 2, 193–194
degenerate, 197–199, 210
nonplanar, 208
non-planar, 209
of octahedra, 199, 210
planar, 198
polyhedral, see Polyhedral isomerizations
of tetrahedra, 198
Isomorphism, 171
Isopectral graphs, 7

Isopolyoxometalates, 161
Itinerant electron density, 145

J

Jahn-Teller-like distortion, 127

K

Keggin structure, see Cuboctahedra
Kuratowski's theorem, 5, 61

L

Lanthanide halide clusters, 148
Lanthanide monohalides, 142
Lanthanide rhodium borides, 153, 169
Lanthanides, 144, 148, 165, 170, 179
Laser vaporization, 125–126, 129
Latent electrons, 117
Latitude, 67
LCAO, see Linear combination of atomic orbitals
Lead, 125, 126, 129
Leapfrog sequence, 190
Ligand labels, 28
Ligand partition, 33
Ligands, 28, 38, 82, 116, see also specific types
Light atoms, 50, 86, 98, see also specific types
Linear axis, 21
Linear combination of atomic orbitals (LCAO), 47, 79
Linear framework groups, 33
Linearly independent wave functions, 36
Linear objects, 21
Linear point groups, 21–22
Line graphs, 4
Liquid insolubility, 144
Liquid nitrogen temperature, 146
Lithium, 131, 178, 181, 183
Localization islands, 86
Localized chemical bonding, 47, 52, 86, 134, 149
Localized systems, 167, see also specific types
Lone electron pair, 82
Longitude, 67
Loops, 8

M

Mackay icosahedra, 185, 186
Macrocuboctahedra, 164, 167
Macrocyclic systems, 162
Macrooctahedra, 164
Macropolyhedra, 162, 163, 165
Magic number, 184
Magnesium, 178
Magnetic behavior, 159
Magnetic fields, 149, 153
Magnetic induction, 146
Magnetic orbitals, 160, 161
Magnetic quantum number, 36
Magnetic susceptibility, 145
Mass spectra, 187
Mass spectrometry, 126
Matching rule, 85, 86
Matrices, 23, see also specific types
- adjacency, 6, 49, 50, 54, 71
- block-factored, 23
- character of, 24
- energy, 49
- overlap, 49

Meissner effect, 146
Mercury, 136–137, 146
Metacyclic groups, 30
Metal borides, 173–179, 181
Metal carbonyl clusters, 86, see also specific types
- alternative theories for, 84–85
- electron counting in, see under Electron counting
- electron-poor, 85–87
- with fused octahedra, 105–110
- with interstitial atoms, see Interstitial atoms

Metal carbonyl vertices, 81–83
Metal clusters, 37, 51, 56, see also specific types
- bare, 125–130
- carbonyl, see Metal carbonyl clusters
- centered, 116–123
- chemistry of, 64
- condensed-phase, 130
- defined, 81
- edge-localized, 88
- electron-poor transition, 86
- gallium, 130–135
- gas phase post-transition, 130
- globally delocalized, 92
- halide, 113–117
- high nuclearity, 105
- indium, 130–135
- infinite, 140
- mercury vertices in, 136–137
- metal carbonyl vertices in, 81–83
- octahedral, 139–145

post-transition, see Post-transition metal clusters
skeletal bonding theories of, 92
skeletal chemical bonding topology of, 139
transition, 113–117, 130, 197
Metal complexes, 28, see also specific types
Metal-metal bonding, 81, 85, 145, 146
superconductors with, see under Superconductors
Metal-metal interactions, 77
indirect, 159
infinite solid state structures with, see Infinite solid state structures
interoctahedral, 148
metal oxides with, see Metal oxides
through oxygen atoms, 159–161
superconductors with, 148
Metal octahedra, 105
Metal oxides, 159–170, see also specific types
aromaticity and, 161–168
delocalization and, 161–168
oxygen atoms and, 159–161
superconducting, 168–170
Metal oxide superconductors, 147
Metals, see also specific types
alkali, 125, 130, 131, 134, 136–137, 181
bulk, 139–145
clusters of, see Metal clusters
coupled, 160
electropositive, 178, 179, 184
middle transition, 42
networks of, 81
pseudo-one-dimensional, 152
transition, 62, 64
weakly interacting, 159
Metal skeletons, 81
Metal tricarbonyl vertices, 82
Metal valence atomic orbitals, 38
Methane, 38, 41
Method of moments, 140
Methylsilyl groups, 83
Metric distortion, 14
Metric features of polyhedra, 14
Middle transition metals, 42, see also specific types
Mismatch, 89
Missing antibonding cluster orbitals, 85
Mixed oxidation, 161
Mixed valence blue, 162
Mixed valence compounds, 167
Mixing, 68, 72, 75–79, see also specific types
Mobile Cooper electrons, 147
Mobile electrons, 147
Molecular bare post-transition metal clusters, 125–130
Molecular beam experiments, 125, 128
Molecular orbitals, 47, 49
ab initio computations for, 73
antibonding, see Antibonding orbitals
binodal, 67
bonding, see Bonding orbitals
cluster valence, 84–85
core, see Core orbitals
degenerate, 64, 72
energy of, 50, 71, 72, 78
external, see External orbitals
Gaussian, 79, 80
Huckel, 49
interactions among, 48
internal, see Internal orbitals
metal oxides and, 165
orthogonal, 36, 65
parameters of, 71
principal core, 72
pure surface, 72, 73, 76
radial, 53
self-consistent computations for, 73, 79
skeletal bonding, 69, 90
Slater-type, 79
square pyramidally hybridized, 83
submolecular, 64
surface, 73, 76, 79, 117
symmetric core, 68
tangential, 53
tetrahedrally hybridized, 83
theory of, 47
trinodal, 67
uninodal, 53, 61, 67, 165
vacant, 197
valence, see Valence orbitals
Molecular skeleton, 49
Molybdenum, 113, 114, 143, 145, 147, 150
infinite solid state structures and, 152
metal oxides and, 159, 165
Molybdenum blues, 161
Molybdenum chalcogenides, 148
Molybdenum dichloride, 114
Molybdenum octahedra, 151
Monocapped tetrahedra, 86
Monohalides, 141, 142
Mononuclear species, 162
Monovalent groups, 83
Muliken symbols, 24–25
Multicenter bonding, 90, 109, 131, 133, 139, 147

Multiple axes, 21
Multiplication, 18, 19, 22, 23, 48, 172

N

Naphthalene, 105, 108
  anthracene and, 108
  edge-sharing, 106–107
  face-sharing, 107
  fusion of, 109
  one-electron reduction of, 162
  superconductors and, 150–151
Neighborhood relationships, 1, 2, 14, 47, 49
Neon, 51
Nickel, 100, 102, 104
Nido structures, 61, 99, see also specific types
  bonding of, 184
  borane, 86
  polyhedral, 61–64, 102, 128
  square pyramid, 129
Niobium, 113, 114, 145, 147
Nitrogen, 115, 116
NMR, see Nuclear magnetic resonance
Nodes, 36, 64, 67, 163, see also specific types
Nonbonding external orbitals, 82
Non-deltahedra, 98
Non-intersecting straight lines, 14
Non-linear framework groups, 33
Non-linear objects, 21
Non-oxide superconductors, 153
Non-planar graphs, 5
Non-planar isomerizations, 208, 209
Non-pyramids, 62, see also specific types
Non-reducible polyoxometalates, 164
Non-spherical valence orbitals, 118–123
Non-triangular faces, 99
Non-trivial symmetry point groups, 25
Normal subgroups, 20, 25, 26, 30
Normal vertex atoms, 50
Nuclear magnetic resonance (NMR), 92, 197

O

Oblate trigonal antiprisms, 174
Octahedra, 4, 12, 13, 15, 16, 20, 42
  bicapped, 95
  capped, see Capped octahedra
  chains of, 152
  complete graphs and, 108
  connectivity of, 200
  core bonding of, 141
  edge-localized, 113–117, 149
  edges of, 57, 74
  electron counting and, 90–92
  electron-rich, 63
  face-fused, 152
  face-linked, 152
  face-localized, 113–117
  face-sharing, 108
  fused, 105–110, 142, 150
  fused pair of, 107
  fusion of, 139–145, 149
  globally delocalized, 113–117
  icosahedral decomposition into, 165
  infinite solid state structures and, 140, 148
  interstitial atoms and, 99
  irreducible representations of, 43
  isomerization of, 199, 210
  metal, 105
  metal carbonyl vertices and, 81
  metal cluster, 139–145
  metal oxides and, 164, 165
  microscopic models of, 196
  molybdenum, 151
  orbital representations for, 73
  polyhedral dynamics and, 200, 202
  polyhedral fragments from, 62
  regular, 14, 34, 92, 163, 207
  1-skeleton of, 59
  spectra of, 59, 75
  structure of, 61
  symmetry point groups of, 58
  tetracapped, 95, 96
  tricapped, 95
Octahedral arrays, 58
Octahedral boranes, 73–80
Octahedral cavities, 141–143, 147, 151, 152
Octahedral core bonding, 141
Octahedral core orbitals, 75
Octahedral graphs, 59
Octahedral groups, 21
Octahedral topology, 58, 59
Odd graphs, 199
Odd irreducible representations, 45
Odd permutations, 28
Omnicapped cubes, 97
Omnicapped dodecahedra, 182
Omnicapping, 190
One-dimensional representations, 24–25
One-dimensional rotation axis, 32
Open polyhedra, 131
Orbitals, 27, see also specific types
  aromaticity of, 161–168
  atomic, see Atomic orbitals
  d, 81
  degenerate, 64, 72
  hybrid, see Hybrid orbitals

magnetic, 160, 161
missing antibonding cluster, 85
molecular, see Molecular orbitals
non-spherical valence, 118–123
orthogonal, 36, 37, 65, 119
p, 81
s, 81
skeletal, 150, 151
submolecular, 65
surface, 73, 76, 79, 117
symmetry of, 197
uninodal, 53, 61, 67, 165
valence, see Valence orbitals
Orthogonality, 161
Orthogonal nature of atomic orbitals, 37
Orthogonal nodes, 163
Orthogonal orbitals, 36, 37, 65, 119
Ortho interactions, 77, 78
Osmium, 88, 89, 93
Overlap integrals, 48, 71, 160
Overlap matrices, 49
Overlap topology, 120
Oxidation, 161, 170
Oxygen atoms, 159–162, 163, 165

P

Para interactions, 76–78
Parallelograms, 136
Paramagnetic transition metals, 159
Parity of permutations, 28
Partial oxidation, 170
Partitions, 29, 33
Penrose tiling, 180
Pentagonal antiprisms, 102–103
Pentagonal bipyramids, 12, 44, 62, 63, 197
belts of, 85
irreducible representations of, 43
microscopic models of, 196
orbital representations for, 73
polyhedral dynamics and, 201
post-transition metal clusters and, 135
Pentagonal faces, 85, 183, 185, 186, 188, 200
Pentagonal pyramids, 12, 16, 42, 44, 63, 175, 177
electron-rich, 64
intermediates of, 210
irreducible representations of, 43
polyhedral dynamics and, 210
quasicrystals and, 184
Pentagons, 3, 34, 187, 189
Pentalene units, 187
Perinaphthene, 108–110
Period of symmetry operations, 19
Peripheral polyhedra, 136
Permutational isomers, 194, 198
Permutation groups, 27–31
conjugacy classes of, 172, 173
cycle structures of, 29
even, 28, 171
identity, 27, 28, 193
odd, 28
parity of, 28
symmetric, 58, 172
transitive, 27, 31, 171
Perspective drawings, 13, 14
Petersen's graphs, 199
2-Phenylbutadiene, 7
Pherical polar coordinates, 35
Phonons, 146
Phopshorus valence electrons, 83
Phosphines, 83, 119, 121, 125
Phosphorus, 83, 91
Pivot face, 195
Planar carbon networks, 187
Planar framework groups, 33
Planar graphite, 179
Planar graphs, 4–5, 15
Planar hexagons, 34, 150
Planar hydrocarbons, 125, see also specific types
aromatic, 34, 47, 54, 71, 163
polygonal, 62, 65, 71
Planar isomerizations, 198
Planar pentagons, 34
Planar polygons, 34, 52, 62
Planar polyhedral isomerization, 193
Planar rectangles, 34, 39
Planar squares, 34, 39
Planck's constant, 35
Point groups, 21–22, 68, see also specific types
symmetry, see Symmetry point groups
Polar coordinates, 35, 36
Polarization, 34
Polgonal hydrocarbons, 62, 65, 71
Polyacene, 151
Polycyclic hydrocarbons, 105, 150
Polygonal aromatic systems, 56, see also specific types
Polygonal graphs, 3
Polygonal hydrocarbons, 51, 53, 62, 65, 81
Polygonal vertices, 53
Polygons, 50, 53
edge-localized, 51
globally delocalized, 51, 60
planar, 34, 52, 62
regular, 3, 4
Polyhedra, 1, 20, 187, 188, see also Pyramids; specific types
arachno, 62, 63, 66, 101, 102

borane, 125
capped, 12, 87
carborane, 125
chiral, 194
cofaces of, 207
combinatorially distinct, 15
coordination, see Coordination polyhedra
diameter of, 209, 210
dual, 13, 182
dynamics of, see Polyhedral dynamics
edge-localized, 51, 134
electron counts for, 84
electron-poor, 87
electron-rich, 61–66, 85, 86
enumeration of, 15
excision of, 62, 63, 87
forbidden, 26, 45, 196, 202
fused, 108
gallium, 134
globally delocalized, 51, 60, 135
gold, 119, 121, 123
Goldberg, 189, 190
hypho, 66
intermediate, 193, 197
isomerizations of, 193–194
metric features of, 14
networks of, 61
nido, 61–64, 102, 128
open, 131
peripheral, 136
properties of, 16
puncture of, 61, 65, 87
pyramidal *nido,* 61
rearrangements of, 98, 195, 204
regular, 21
simple, 52
skeletal bonding within, 50
1-skeleton of, 61
symmetry of, 187, 188, 201
three-dimensional, 61
topologically distinct, 15
topology of, 10–16
vertices of, 32
Polyhedral boranes, 37, 51, 52, 125
Polyhedral boron, 130
Polyhedral carboranes, 51, 52, 125
Polyhedral cavities, 105, 184
Polyhedral dynamics, 193–210
gale diagrams and transformations and, 204–210
isomerizations and, 193–194
macroscopic models of, 193, 194, 198–203
microscopic models of, 193–198
Polyhedral gallium, 130–135
Polyhedral groups, 21–22, see also Polyhedra
Polyhedral indium, 130–135
Polyhedral isomerizations, 210
degenerate, 193, 197–199, 210
nonplanar, 208
non-planar, 209
planar, 193, 198
topological representations of, 198–203
Polyhedral point groups, 21–22
Polyhedral skeletal electron pair theory, 84
Polyhedral skeleton, 28
Polyhedral vertex labels, 28
Polyhedranes, 52
Polynomials, 6, 9, 10
Polynuclear transition metals, 159, see also specific types
Polyoxometalates, 159, 162, 165–167
early transition metal, 161–168
non-reducible, 164
Polyoxotungstate, 162
Polytopes, 204, 205
P orbitals, 81
Porcupine compounds, 120
Porous conducting skeleton, 158
Porous delocalization, 149
Porous infinite delocalization, 153
Post-transition elements, 42, 83, see also specific types
Post-transition metal clusters, 125–137, see also specific types
gallium, 130–135
gas phase, 128, 129
indium, 130–135
ionic bare, 125–130
mercury vertices in, 136–137
molecular bare, 125–130
Potassium, 136, 182
Potential energy, 35
Primary involution, 26, 45
Primes, 25
Primitive root, 31
Principal core orbitals, 56, 57, 72, 78
Principal eigenvalue, 56, 165
Principal quantum number, 365
Prisms, 43– 46, see also specific types
tricapped trigonal, see Tricapped trigonal prisms
trigonal, see Trigonal prisms
Proper rotations, 26, 173, 194
Proton chemical shift, 92
Pseudo-deltahedra, 101, 105
Pseudohalide, 125
Pseudohalogen, 121
Pseudohexagonal faces, 199
Pseudo-one-dimensional metals, 152
Pseudoreflection, 198
Pseudorotation, 196, 197, 209
Pure bonding surface orbitals, 79

Pure complete topology, 58
Pure surface orbitals, 72, 73, 76
Pyramidal *nido* polyhedra, 61
Pyramids, 62, 64, 65, 85, see also specific types
  base of, 208
  heptagonal, 201
  hexagonal, 64
  pentagonal, see Pentagonal pyramids
  square, see Square pyramids

Q

Quadrilateral circuits, 8
Quadrilateral faces, 11, 98, 101, 195, 210
Quantum numbers, 36
Quarternary ammonium amalgams, 136
Quasicrystals, 173, 180–186

R

Radial component of wave functions, 36
Radial molecular orbitals, 53
Radial vectors, 67
Radical anions, 162
Reciprocals, 19
Rectangles, 34, 39, 136
Reducible representations, 23, 45
Reduction, 167
Reference axis, 22
Reference configurations, 28
Reference cubes, 157
Reference reactants (chemical species), 2
Reflection, 17, 22, 23
Reflection operation, 18
Reflection planes, 10, 17, 18, 27, 32, 45
Regular dodecahedra, 52, 188, 189, 198
Regular graphs, 7
Regular heptagons, 34
Regular icosahedra, 62, 79
Regular octahedra, 14, 34, 92, 163, 207
Regular polygons, 3, 4
Regular polyhedra, 21
Regular tetrahedra, 34, 38, 171
Relativistic effects, 119
Representation of groups, 23–25
Resistive heating, 125, 129
Resonance hybrids, 163
Resonance integrals, 71
Resonance stabilization, 55, 100, 187
Resonating valence bonding, 170
Rhenium, 147
Rhodium, 89, 90, 105, 108, 110, 154
  vertices of, 90, 107, 109
Rhodium borides, 153
Rhodium carbonyl anions, 148, 150
Rhodium clusters, 150, 153
Rhodium-rhodium bond, 106
Rhombohedra, 174
Rhombohedral axes, 174
Rhombohedral boron, 175, 176, 178, 183
Ring current, 92
Rotation axis, 10, 17, 18, 21, 25, 32
Rotations, 17, 22
  improper, 17, 25, 32–33, 173
  proper, 173, 194
  pseudo-, 196, 197, 209
  subgroups of, 171
Ruthenium, 106

S

Sachs graphs (sesquivalent subgraphs), 7, 8
Sampling errors, 72
Samson complexes, 175, 177, 183, 184, 186, 187
Scalar spherical harmonics, 67
Schlegel diagrams, 13–15
Schoenflies symbol, 20, 32
Schrodinger equations, 35, 67
S-d hypothesis, 145
Second-order differential equations, 35, 67
Secular equations, 48
Selenium, 126, 151, 152
Self-consistent molecular orbital computations, 73, 79
Semiconducting energy gap, 141
Semiconductors, 144, 167
Semidirect products, 25, 26, 171, 172
Series orthogonality, 161
Sesquivalent subgraphs (Sachs graphs), 7, 8
Seven-vertex systems, 127
Silicon, 156
Silverton structure, 164, 165
Similarity transform, 20, 23
Simple groups, 20
Simple polyhedra, 52
Simultaneous equations, 48
Single interior vertex atoms, 65
Site partition, 33
Skeletal bonding, 49, 66, 67, 81
  cluster, 83, 84–85
  edge-localized, 100
  electron counting and, 84–85
  of icosahedra, 175
  icosahedra and, 176
  infinite solid state structures and, 156
  intrapolyhedral, 131
  metal carbonyl cluster, 84–85
  orbitals of, 69, 90
  within polyhedra, 50
  theories of, 92
  topology of, 139, 156
Skeletal electrons, 51, 82, 83, 86, 89, 92

apparent, see Apparent skeletal electrons
biphenyl analogue and, 106
capped octahedra and, 94
counting of, see Electron counting
early transition metal clusters and, 115, 119
fused octahedra and, 107
globally delocalized octahedra and, 113
icosahedra and, 184
infinite solid state structures and, 139, 141
interstitial atoms and, 100, 101, 102
net, 115, 116
octahedra and, 90, 107, 113
post-transition metal clusters and, 126, 130, 134, 135
sources of, 110, 150, 151
total, 135, 141
uses of, 110
Skeletal orbital sources, 150, 151
Skeleton, see also specific types
allene (bisphenoid), 34
conducting, 148, 156, 158, 168, 169
metal, 81
molecular, 49
polyhedral, 28
porous conducting, 158
1-Skeleton, 1, 11, 54, 56, 59, 61, 149
polyhedral dynamics and, 202
Slater functions, 79
Slater-type orbitals, 79
Sodium, 136, 178
Solid solubility, 144
Solid state structures, infinite, see Infinite solid state structures
Solubility, 144
Soluble groups, 20
S orbitals, 81
Spd-forbidden corrdination polyhedra, 45
Spheres, 15, 67, 187
Spherical harmonics, 35, 36, 67
Spherically symmetric atomic orbitals, 37
Spherical polar coordinates, 36
Square antiprisms, 44–46, 100–101
bicapped, 102, 128, 196
capped, 43, 46, 102, 114, 127, 129–130
intermediates of, 203
irreducible representations of, 43
polyhedral dynamics and, 201, 202
post-transition metal clusters and, 128
symmetry of, 201
Square faces, 14
Square planar complexes, 118
Square planar intermediates, 193, 198
Square pyramidally hybridized orbitals, 83
Square pyramids, 14, 16, 42, 62
electron-rich, 63, 64
intermediates of, 198
irreducible representations of, 43
nido, 129
symmetry of, 197
Squares, 3, 9
mercury, 136, 137
planar, 34, 39
polyhedral dynamics and, 202
post-transition metal clusters and, 127, 137
Square of symmetry operations, 18
Stabilization, 55, 59, 60, 100, 187
Stabilizers, 27
Stoichiometry, 56, 134, 156, 184
Strain, 187
Structural stability index, 162
Sturred vertices, 183
Subgraphs, 3, 6, 7, 8
Subgroups, 19, 20, 25, 26, 30, 171, see also specific types
Submolecular orbitals, 64, 65
Sulfur, 83, 148–150
Sum of squares, 24
Superconducting copper oxides, 168–170
Superconductors, see also specific types
bismuth copper oxide, 168
chemical bonding topology of, 153, 158
copper oxide, 146, 159, 168–170
defined, 146
high critical temperature copper oxide, 159
with metal-metal bonding, 146–158
anthracene and, 151
copper oxide, 146
defined, 146
metal oxide, 147
naphthalene and, 150–151
non-oxide, 153
temary, 148
with metal-metal interactions, 148
metal oxide, 147
non-oxide, 153
tenary, 148
ternary, 156
thallium copper oxide, 168
Superexchange, 159
Surface bonding, 108, 110, 113
early transition metal clusters and, 119
of icosahedra, 173, 175
icosahedra and, 176, 177
post-transition metal clusters and, 135
Surface current, 92
Surface energy units, 72
Surface harmonics, 37, 56, 67–69, 115, 117
Surface holes, 99

Surface orbitals, 73, 76, 79, 117
Switching edge, 195, 196
Symmetric core orbitals, 68
Symmetric groups, see Symmetry groups
Symmetric irreducible representations, 25, 26, 56
Symmetric permutation groups, 58, 172
Symmetry, 10, 36, 41, 198
- of bisdisphenoid, 201
- of cubes, 201
- defined, 17
- of hexagonal bipyramids, 201
- horizontal plane of, 22, 25
- of icosahedra, 171–173, 182
- macropolyhedral, 163
- orbital, 197
- of polyhedra, 187, 188, 201
- special, 72
- of square antiprisms, 201
- of square pyramids, 197
- of two-dimensional simplex, 171
- vertical planes of, 22

Symmetry elements, 14, 17, 201
Symmetry factoring methods, 10
Symmetry groups, 27, 28, 33, 79, 201
Symmetry operations, 17–18, 36
Symmetry plane, 25
Symmetry point groups, 18–23, 32, 58
- defined, 18, 20
- as direct product groups, 25–26
- direct product structures of, 27
- of icosahedra, 171
- irreducible representations of, 24–25, 39
- non-trivial, 25
- of octahedra, 58
- polyhedral dynamics and, 194

Synmmetry elements, 10

T

Tangential orbitals, 53
Tantalum, 147
Technetium, 147
Tellurium, 42, 126, 151, 152
Temary superconductors, 148
Tensor surface harmonic theory, 37, 56, 67–69, 115, 117
Teo theory, 84
Ternary lanthanide rhodium borides, 153
Ternary superconductors, 148, 156
Tertiary phosphines, 83, 119, 121, 125
Tesseracts, 202
Tetracapped octahedra, 95, 96
Tetracapped tetrahedra, 153, 154
Tetracoordinate boron, 177
Tetragonal antiprisms, 100–101
Tetragonal bipyramids, 85
Tetragonal forms, 174
Tetrahedra, 1, 3, 4, 16, 20, 31, 39
- bicapped, 16, 86, 87, 92–93
- capped, 86
- edge-localized, 113–117, 153
- electron counting and, 81, 85, 87–88
- fused, 86, 92
- infinite solid state structures and, 153, 154
- interconversion of, 193
- irreducible representations of, 43
- isolated, 86
- isomerization of, 198
- microscopic models of, 196
- monocapped, 86
- post-transition metal clusters and, 127
- regular, 34, 38, 171
- tetracapped, 153, 154
- vertex atoms and, 52
- vertices of, 41

Tetrahedral cavities, 140, 142, 143
Tetrahedral chambers, 52, 86, 87, 93, 96, 141
- fused, 86
- polyhedral dynamics and, 196
- post-transition metal clusters and, 129, 135

Tetrahedral groups, 21
Tetrahedrally hybridized orbitals, 83
Tetrahedral rhodium clusters, 153
Tetrahedral rotation groups, 31
Thallium, 126, 130, 169, 170
Thallium copper oxide superconductors, 168
Three-center face bonds, 114, 115
Three-dimensional aromaticity, 55, 56, 126
Three-dimensional deltahedra, 63
Three-dimensional polyhedra, 61
Three-dimensional representations, 24
Three-dimensional simplex, 171
Three-fold synmmetry elements, 10
Tiling, 180
Tilting, 148
Tin, 103, 125, 126, 129
Topological homeomorphism, 67
Topologically distinct polyhedra, 15
Topological representations, 2, 194
Topological spaces, 1, 2
Topology, 1–3, 47
- of chemical bonding, see Chemical bonding topology
- complete, 58, 59, 76
- of complete graphs, 90
- core bonding, 58, 76
- defined, 1

deltahedral, 56, 59
octahedral, 58, 59
overlap, 120
of polyhedra, 10–16
pure, 58
of skeletal bonding, 139, 156
Torus, 118, 187
Totality of faces, 11
Totality of vertices, 12
Trans interactions, 58, 74
Transition metal clusters, 113–117, 130, 139, 197, see also specific types
Transition metal divide, 144, 147
Transition metals, 62, 64, 98, 118, 159, see also specific types
Transition metal vertices, 81, 83, 84, 86
Transitive framework groups, 34
Transitive groups, 30
Transitive permutation groups, 27, 31, 171
Transposition, 28
Triangles, 3, 8, 34, 81, 129, 150, 171
Triangular faces, 11, 12, 15, 62, 87, 98, 108
polyhedral dynamics and, 195
Tricapped octahedra, 95
Tricapped trigonal prisms, 46, 107, 142
electron counting and, 96
irreducible representations of, 43
orbital representations for, 73
polyhedral dynamics and, 196
post-transition metal clusters and, 127–128
Tricarbonyl vertices, 82
Trigonal antiprisms, 34, 174
Trigonal bipyramidal carboranes, 197
Trigonal bipyramids, 16, 22, 32, 86, 87
belts of, 85
coordination numbers and, 42
electron counting and, 88–90
elongation of, 85
homonuclear, 90
irreducible representations of, 43
microscopic models of, 196
polyhedral dynamics and, 195, 198, 199
post-transition metal clusters and, 127, 130
Trigonal prisms, 16, 34, 44–46
bicapped, 43, 45, 46
capped, 43, 44
connectivity of, 200
delocalized, 96
globally delocalized, 96
interstitial atoms and, 100
irreducible representations of, 43
isomers of, 200
polyhedral dynamics and, 200
tricapped, see Tricapped trigonal prisms
Trimethylsilyl groups, 83
Trinodal orbitals, 67
Triple dehydrogenation, 100
Tris(ethylenedithiolate), 42
Trivalent phosphorus ligands, 82
Tropylium, 82
Truncated icosahedra, 175, 182, 183, 187, 189
Tungsten, 143–145, 147, 159
Tungsten blues, 161
Two-center chemical bonds, 47
Two-dimensional polygonal aromatic systems, 56
Two-dimensional representations, 24
Two-dimensional simplex, 171
Two-electron exchange integral, 160
Two-electron multicenter bonds, 139
Twofold symmetry element, 10
Two-particle interactions, 146
Type I structures, 163, 164, see also specific types
Type II structures, 164, see also specific types

U

Ungerade irreducible representations, 40
Unidentate ligands, 38
Uninodal orbitals, 53, 61, 67, 165

V

Vacant molecular orbitals, 197
Valence atomic orbitals, 38
Valence bands, 148, 158
Valence bonding, 47, 160, 170
Valence electrons, 83, 126
Valence orbitals, 81, 126
atomic, 38
cluster, 84–85
icosahedra and, 176
infinite solid state structures and, 143
non-spherical, 118–123
vandium, 157
Vanadium, 145, 156, 157
Vandium, 147, 156
Vandium valence orbitals, 157
Vector operators, 68
Vector surface harmonics, 68
Vertex atoms, 64, 82
border, 62
delocalization and, 50–52
heavy, 82

interior, 62, 65
internal orbitals of, 67
orbitals of, 139
single interior, 65
Vertical edges, 196
Vertical planes of symmetry, 22
Vertices, 1, 2, 3, 188, 198, see also specific types
alkylarsinidene, 96
alkylphosphinidene, 83, 91
allowed motions of, 204, 208–210
antipodal, 74, 102
apical, 62, 88
axial, 22, 89, 196
bare phosphorus, 83
basal, 62
boron, 86, 131
carbon, 86
of degree, 11
edge relationship with, 11
equatorial, 22, 89, 174, 196, 198
gallium, 130
hypervalent, 83, 91, 96
mercury, 136–137
metal carbonyl, 81–83
metal tricarbonyl, 82
neighborhood relationships of, 14
osmium, 93
phosphorus, 83
polygonal, 53
of polyhedra, 32
rhodium, 90, 107, 109
sturred, 183
totality of, 12
transition metal, 81, 83, 84, 86
tricarbonyl, 82
Volume, 105

W

Wade-Mingos theory, 84
Walk of length, 5
Wave functions, 35–37, 48, 149, 160
Weakly interacting metals, 159
Wreath product groups, 201

X

X-ray crystallography, 125
X-ray diffraction, 90, 91, 125, 126

Y

Yttrium, 179

Z

Zero point of energy, 71
Zinc, 181
Zintl phases, 125
Zirconium, 115, 116, 142
Zirconium monohalides, 141
Zone of influence, 144